PACIFIC VOICES: EQUITY AND SUSTAINABILITY IN PACIFIC ISLANDS FISHERIES

Editors:

Irené Novaczek
Institute of Island Studies
University of Prince Edward Island
Charlottetown, Canada

Jean Mitchell
Dept. Sociology and Anthropology
University of Prince Edward Island
Charlottetown, Canada

Joeli Veitayaki
Marine Studies Programme
University of the South Pacific
Suva, Fiji

Institute of Pacific Studies Publications
2005

USP Library Cataloguing-in-Publication Data

Pacific voices : equity and sustainability in Pacific Islands fisheries / editors, Irené Novaczek, Jean Mitchell, Joeli Veitayaki. – Suva, Fiji : Institute of Pacific Studies, University of the South Pacific, 2005.

246 p. ; 18 x 25 cm.

ISBN 982-02-0372-4

1. Sustainable fisheries—Oceania 2. Fishery management—Oceania 3. Fisheries—Oceania—Case studies 4. Shellfish fisheries—Oceania 5. Women fishers—Oceania I. Novaczek, Irené II. Mitchell, Jean III. Veitayaki, Joeli IV. The University of the South Pacific. Institute of Pacific Studies.

SH329.S87P23 2005 333.9560995

Cover design: Pro Photos Fiji
Book design & typesetting: Mark Garrett
Indexing: Irené Novaczek
Proofreading: Godfrey Baldacchino & Joeli Veitayaki
Production: Maria Ronna Luna Pastorizo & Mark Garrett
Printing: Oceania Printers Limited, Suva, Fiji Islands

Institute of Pacific Studies, University of the South Pacific, PO Box 1168, Suva, Fiji, phone +679 321 2018, fax +330 1594, e-mail ips@usp.ac.fj
online book sales http://www.pacificstudies.org

PACIFIC VOICES

Contents

FOREWORD

This book is a welcome addition to the slowly growing and compelling literature on gender and inshore fisheries in the Pacific. It brings together a number of papers that range across old and new issues related to the development of commercial and export fisheries, which men still largely control, and the survival of subsistence fisheries, in which women predominate. In a seemingly neat gender division of resource bases and labour, Pacific Islands women exploit inshore waters to provide food for family consumption, while men prosecute deep-sea fisheries to produce commodities for domestic sale and export. Although subsistence fisheries provide food security for many thousands of Pacific Islanders, the labour of women is not reflected in national accounts. By contrast, foreign exchange earnings from export fisheries are reflected in national accounts and are seen to hold the promise of national economic growth.

The papers in this volume highlight the intersections between commercial and subsistence fisheries. They explore the economic imperatives and the social, cultural and gender issues attached to these fisheries. In highlighting these intersections, the papers hint at the tensions and contradictions at play in coastal communities. On one hand is market oriented production and the private profit motive that drives it. On the other hand there is subsistence production and the principles of sufficiency and conservation that are embedded in it. Such tensions are not peculiar to Pacific Islands societies. However, our region's retention of communal-based resource ownership systems makes the subsistence option a viable alternative for large numbers of people in resource-owning groups. Elsewhere in the world, the privatization of 'the commons' has largely removed that option, and the dispossession entailed has brought not only dependence on the market economy, but also economic impoverishment.

The subsistence sector and the inshore fishery are at risk in the Pacific region, not least because market imperatives hold sway. Moreover, prevailing gender systems in the Pacific work to exclude key stakeholders in the inshore fishery from contributing to resource use decision-making. This is detrimental to the inshore fishery and, ultimately, to the subsistence way of life itself. The interface of economic development and gender systems has, ironically, left women largely responsible for the subsistence sector while withholding the political power to protect this important turf. A key challenge for women fishers and for their advocates lies in demanding that women be given central positions in decision-making around the use of inshore fisheries resources.

Tensions between market imperatives and the maintenance of an old way of life are intensifying under pressures of economic globalization and trade liberalization. In a free-for-all world of global free trade, the strongest players in the market will take all. For example, the plundering of the Pacific's tuna and other deep sea resources has been going on for years under licensing arrangements that Pacific Islands states have with Distant Water Fishing Nations. These nations have the technology and financial resources to 'efficiently' (in fact, wastefully) mine our seas in search of species that fetch the highest market price. Whole fisheries elsewhere in the world (e.g. the North Atlantic cod) have collapsed from precisely the same high-tech, profit-driven, plundering approach. The paltry earnings that Pacific Islands states receive from 'fishing rent' (averaging 3-4% of the value of the catch) contrast with the multi-millions netted by global fishing companies each year from the massive extractions of fish from the Pacific Ocean. Our concern, however, should be less with how much we receive in financial returns from this wasteful over-extraction of resources for private profit, than with how much fish we should allow to be taken from our seas each year.

At the national level, deciding who should have access to fisheries resources and on what terms can become a matter of fierce contestation, rather than agreement based on considerations of social equity and resource sustainability. In 2003, for example, the Fisheries Department in Fiji issued a large number of licences to Chinese companies that had developed joint venture tuna fishing partnerships with indigenous entities. This triggered a major reaction from the country's leading tuna exporter. In the ensuing debate with Fisheries Department, representatives of the private vested interests were effectively juxtaposed with an implicit indigenous rights claim. The question of whether the fishery could sustain more intensive exploitation became a moot point.

As some of the papers in this volume show, the decline of inshore fish stocks and habitats is not unrelated to export fisheries. Industrial bait fishing has long been known to threaten the inshore fishery, as have the live fish and live coral trades, and the intensive exploitation of other niche market export products such as bêche de mer. The social dimensions and impacts of these industries have been less studied until now. Likewise, the attention in this volume to the economic, social and gender dimensions of ornamental shellfisheries in Solomon Islands and Tuvalu brings new knowledge to the field.

Building understanding of fisheries conservation and management issues within our communities - global, regional and local - especially in relation to inshore fisheries, is vital. Equally important is building understanding of the serious impediments to sustainable development that are posed by inequitable gender systems. We need more research and analysis of what is happening in inshore fisheries today. We need wide publication of the results of such research and of the analyses produced by biological and social scientists. Finally, we need more community workshops with fisherwomen to discuss these issues and others that affect them.

This collection makes a valuable contribution to knowledge about the inshore fisheries and gender issues. It is hoped that the publication will have multiple uses: in community and public education; in advocacy for a more gender-equitable and responsible approach to fisheries development; in the struggle for gender equality more generally; and in stimulating further research work on a vital resource sector in the Pacific Islands.

Claire Slatter
Founding member, Pacific Women and Fisheries Network
General Coordinator, Development Alternatives with
Women for a New Era (DAWN)

PREFACE: A UNIQUE COLLABORATION

This book is the result of long-term support for Pacific fisheries development by the Canadian Agency for International Development (CIDA). Since 1990, through two phases of the Canada South Pacific Ocean Development Program (C-SPOD), Canada has supported Pacific regional institutions to assist Pacific Islands Countries to manage their ocean resources sustainably. In addition, 35 Pacific Islanders have been supported for postgraduate training in marine issues, at the University of the South Pacific. Gender aspects of fisheries development have received particular attention. C-SPOD was a pioneer in supporting women and fisheries projects, initiating one such program in the early 1990's at the Secretariat of the Pacific Community (SPC). This subsequently evolved into the Community Fisheries Program. Gender issues are now integrated into all C-SPOD sub-projects. Training in gender analysis has been offered to the regional partner institutions.

The second phase of C-SPOD had a specific objective to increase the knowledge on "women's fishing-related activities so that the knowledge becomes more visible, more valued, and more accurately documented". To meet this objective, projects were solicited. As a result, this unique initiative was undertaken jointly by the University of the South Pacific (USP) in Fiji and the University of Prince Edward Island (UPEI) in Canada, with support from the Secretariat of the Pacific Community (SPC). The project assisted Pacific Islands researchers in conducting research and publishing much-needed information on the harvesting, processing, marketing and management of Pacific fisheries resources. Many of these researchers were current or former graduate students under the C-SPOD Program at USP. This project was part of the effort of USP, through its Marine Studies Programme, to work in collaboration with other institutions internationally to promote truly equitable and sustainable development for the small-scale fisheries sector. UPEI, through its Institute of Island Studies, brought to the collaboration its history of research and community work that seeks to reveal and analyse the particular development opportunities and challenges faced by peoples living on small islands.

Dr Irené Novaczek (Project co-ordinator, UPEI), Dr Joeli Veitayaki (Project leader, USP) and Dr Jean Mitchell (Project leader, UPEI) collaborated in the needs assessment, project planning and execution of the project. The initiative commenced with a two-week training workshop in Fiji during June 2003. Sixteen Pacific Islanders worked with Drs Veitayaki and Novaczek and with invited speakers: Aliti Vunisea and Dr Mecki Kronen (SPC), Neil White (Biology Department USP), Margaret Leniston (Forum Secretariat) and Dr Vanessa Griffen. The resource persons provided inspiration, skills and knowledge through seminars and hands-on exercises. The training materials focused on the importance of small-scale and subsistence fisheries, recent progress in community-based management of marine resources and the critical need for a gender-sensitive approach to fisheries research.

For two weeks, workshop participants studied, read, discussed and wrote about how issues of gender, sustainability, modernization and fisheries intersect. They learned about and practised research and analytical methods, examined a gender-sensitive research framework for fisheries, and developed their own work plans and research instruments.

By the end of two weeks of intensive work, each researcher was prepared to return to his or her country to develop a case study on some aspect of fisheries important to the Pacific region. Each case study considers and analyses the roles of women as well as men, with most focusing on the small-scale fisheries that provide food security and livelihoods to many rural Pacific Islanders.

The project provided workshop participants with financial support for their research and a mentor was assigned to each participant. The mentors in some cases followed up with the researcher in the field. The researchers included graduate students, government fisheries and rural development officers, staff of regional organisations and NGO-based community workers. Many chose to work in their home communities and so brought a particular depth of local knowledge to their case studies. The case study writers were from seven Pacific Islands countries. They represent potential leaders for fresh and effective approaches to fisheries development and management in their own communities and the region. We are very proud of their accomplishments and hope that this will be the first of many publications that will help to build a firm foundation for sustainable inshore fisheries of the future.

Kenneth T. MacKay PhD
Field Program Coordinator
Canada-South Pacific Ocean Development Program II
Pacific Islands Forum Secretariat, Suva, Fiji

Jane Ledwell
Director
Institute of Island Studies, University of Prince Edward Island
Charlottetown, Canada

Leon Zann PhD
Director
Marine Studies Programme
University of the South Pacific
Suva, Fiji

1 Voices, Lenses And Paradigms: Understanding Fisheries Development In The Pacific

Joeli Veitayaki and Irené Novaczek

INVISIBLE WOMEN, DISAPPEARING FISH

Two frequently heard commentaries on Pacific fisheries are that inshore fish stocks and habitats are in decline, and the role of women in small scale and subsistence fisheries is relatively neglected. Despite the fact that these concerns have been repeatedly voiced over the past decade, data on the status of inshore stocks and data on the volume of fish and shellfish. caught by small-scale, inshore fishers are still rare or absent for many Pacific Islands nations. Similarly, information on fisheries activities of women, the impact of these activities on fish stocks and habitats and fisherwomen's contributions to the economy, society and culture of Pacific Islands is also missing, incomplete or superficial. In part this is because governments that fund research often assume that the only truly important fisheries are those which earn foreign currency and promote modernization. Yet, the inshore provides essential protein to the majority of Pacific Islanders and in many island nations, women are the key labour force in subsistence fisheries.

Traditional fishing canoe on the beach of Lelepa Island, Vanuatu

If inshore resources critical to regional food security are truly in decline, it is important to document this decline, determine the reasons behind it and look for options to restore, conserve and manage these essential resources. In researching such topics it is essential that a gender lens be employed, that is, that women's as well as men's histories, roles and perceptions are documented. Finally, any such documentation benefits from the involvement of researchers who are rooted in Pacific culture and able to grasp the often unspoken complexities of the relationships between coastal peoples and their inshore marine environments.

INTRODUCTION TO THE CASE STUDIES

The case studies in this book were written to enhance our understanding of coastal fishing communities of the Pacific region at this critical point in their history. In many cases the writers grew up in coastal villages and fished alongside their mothers and fathers as children. These Pacific voices tell, often through direct quotes from village people, what islanders remember of the past riches of inshore fishing grounds, how and why changes have occurred, what they think of current conditions and how they see their prospects for the future.

Women on Abaiang Atoll, Kiribati, preserve fish by salting and drying.

Unlike much fisheries research that focuses on the biology of commercially exploited species or the function of supporting ecosystems, we privilege social issues. In particular, we apply social science methodologies to explore the ways in which culture, gender and generational perspectives interact with the processes of modernization. We seek to understand how this interplay influences the rate and shape of technological change in small-scale fisheries, the intensity of fishing effort, the need for conservation and management, the shape of local management institutions and the success or failure of management strategies. Only by grappling with the interplay of history, custom, economics and social relations can a researcher uncover the incentives and motivations to fish or not fish; to destroy or conserve resources; and to participate in or resist management efforts.

The case studies show that the rate and degree of change in inshore fisheries over the past century have indeed been profound, and many communities now face difficult challenges. We begin with Samasoni Sauni and Lilian Fay-Sauni, who highlight the vulnerability of inshore fisheries and the dependence of Tuvaluans on these resources. Then we take a look at some other traditional fisheries that are struggling to survive and thrive against serious odds. Kristina Fidali-Hickey and Cherie Whippy-Morris give us a glimpse of the shell money industry of Malaita, Solomon Islands where the demands of the modern economy, changing cultural values and social unrest combine to place people and their resources at extreme risk. Alan and Setapu Resture investigate another cultural phenomenon that has evolved into a commercial enterprise: the ornamental shellfishery in Tuvalu. In both cases there is no effective management of shell resources. Women, who are vulnerable because of their limited employment opportunities, are heavily engaged in these shell trades and therefore women are most at risk if shell resources collapse. The lessons to be learned from the Solomon Islands' experience are a warning for Tuvalu. In Solomon Islands, some shell resources have already been driven to local extinction. In Tuvalu, there are opportunities to avert that consequence.

Lilian Fay-Sauni and Samasoni Sauni present evidence of decline in the edible ark shell resources of Tarawa, Kiribati. Among other findings, they note how knowledge of the social habits of the primary fishers, who are women, is important to the design of approaches for engaging fishers in management. Joeli Veitayaki and his son Vesi show us what it is like to participate in a traditional fish drive in Fiji, and consider how such an ancient practice connects with modern fishing methods and management strategies.

The next cluster of case studies focuses on change and conflict. First, Aliti Vunisea looks at changing gender roles in village fisheries in Fiji. She reveals the expertise of the women who are "only gleaning" in shallow waters, accentuates the social and cultural dimensions of women's work and discusses how these evolve into strategies for coping with commercialization and globalization. Alan Agassi reviews the history of change in the fisheries in his home village in Solomon Islands where women and girls have entered the fishery in unprecedented numbers. He also explores social impacts of the industrial bait fishery. Samasoni Sauni and his colleagues at the SPC reflect on the implications of a new kind of fishery, the live coral trade in Fiji. They question whether the economic benefits balance the long term risks to ecology and society. Siola'a Malimali considers the merits and future of yet another Pacific experiment in fisheries development: the introduction of *Trochus*, an exotic species, into Tongan inshore waters.

Next, we present two views of post-harvest activities. First, Jyotishma Rajan illustrates issues of gender and labour in industrial tuna processing in Fiji by focusing on two women's personal stories. Maere Tekanene relates her action research into the situation of women in the informal fish trade of Kiribati. In the case studies focused on village activities, solutions to current challenges are often seen to lie within the community, with or without assistance from outside. For women engaged in post-harvest processing and marketing, the interventions required to improve their conditions range from the local to national and international in scope.

Finally, we take a look at some of the challenges that arise when communities attempt to take control of their destinies and manage their local fisheries. Jean Tarisesei and Irené Novaczek offer insights into the importance of integrating women and youth into fisheries management in rural Vanuatu. Malama Siamomua-Momoemausu offers hope with her account of the involvement of women in fisheries management in Samoa. To bring the volume to a close, Vina Ram-Bidesi and Jean Mitchell provide us with an overview that finds the threads that bind these varied studies with one another and with the Pacific fisheries literature.

River clams (kai) for sale at Suva market, Fiji.

TENSIONS IN FISHERIES DEVELOPMENT: GENDER, GENERATION, MODERNIZATION

Various researchers, especially those of the Secretariat of the Pacific Community's Fisheries Division, the Forum Secretariat and the USP-based Women in Fisheries Network (e.g. Matthews 1995) have stated that women make vital contributions to the functioning of Pacific societies in general and to subsistence fisheries in particular. Women are wholly responsible for certain fisheries-related tasks and often charged with providing food for their families. Young women form the most vigorous labour force in many subsistence fisheries and so are well placed to monitor changes in inshore resources and offer advice on how to manage and conserve these resources. However, although youth under age 20 may make up over 50% of the population, in Pacific societies it is elders, particularly males, who hold decision-making authority. Yet youth cannot be ignored: they face the daunting challenge of reconciling traditional values with an emerging consumer society, the demands of which put significant pressure on fisheries and other natural resources.

Examining the impacts of the development of Pacific Islands fisheries at the community level, we see the tensions between generations and between genders, as well as the complex inter-relationship of tradition and modernity. The introduction of modern technologies, the influence of the church, the growing desire for consumer goods, and the drive to provide education for children have all influenced how and with what intensity people fish. Islanders remain highly dependent on fisheries resources for food and income (Sauni & Fay-Sauni, this volume) and fishing activities are valued expressions of rural culture (Malimali, this volume, Veitayaki & Vesi, this volume). Nevertheless, cash incentives have frequently led to over-harvesting of valued species (Fidali-Hickey & Whippy-Morris, this volume, Fay-Sauni & Sauni, this volume) or to habitat destruction (Sauni et al, this volume). The interplay among technology, culture, custom and the cash economy has also influenced gender roles in rural communities, the social positions of youth and women and perceptions of traditional authority. Whereas in some cases access to modern technologies and commercial markets has empowered women and youth (Agassi, this volume, Vunisea, this volume), in other cases women have been excluded from commercial fisheries or their subsistence harvests have been appropriated by men for commercial exploitation (Tarisesei & Novaczek, this volume). Traditional owners who control fishing grounds may invite development that has particularly negative impacts on women and children (Agassi, this volume). Social change and development may have the unintended result of increasing the work load and financial responsibilities of women (Siamomua-Momoemausu, this volume, Rajan, this volume). At the same time, cultural norms have tended to exclude the voices of women from the decision-making processes that are critical to community well being (Tekanene, this volume, Resture & Resture, this volume).

Because men and women experience development differently, gender analysis is an important tool for understanding fisheries development. Through gender analysis the researcher identifies the roles played by women, men, boys and girls in the household, community, workplace, political process and the economy. Gender analysis explores the different perspectives and interests of men and women and reveals who has access to and control over resources, benefits and decision-making processes. It looks at the potentially different impacts of an intervention on males and females, and uncovers why men and women have different constraints and opportunities. By showcasing gender analysis in this book, we hope to encourage its use in the Pacific so that future coastal developments will be more equitable, will take into consideration the well being of all members of society and will therefore be more sustainable.

DEVELOPMENT PARADIGMS OF THE PACIFIC

It is easy to argue that the need for effective rural development is critical for small Pacific Islands nations that have marginal economies. The islands are small and isolated, with poorly developed infrastructure and uncertain transportation links. Access to education, health services and employment is often limited. In past attempts to overcome such challenges, Pacific Islanders have freely adopted models of development from elsewhere. This has not served the region well.

For decades, women in the developing world have critiqued the dominant development paradigms. Analyses have shown that development through industrialisation, which prevailed for decades, was generally ineffective and in fact made life worse for women and the poor who lived in developing countries. This development process downplayed the role of women as keepers of culture and tradition and the contributions of women to production and reproduction in society. It also failed to recognise the degree to which women rely on indigenous institutions and cultural practices for survival. Indigenous, traditional attitudes and institutions were viewed as barriers to modernization and development. More recently, development practitioners have come to realise that if developing countries are to prosper, women need to be recognized and integrated into the development process. The southern women's network DAWN, which is currently based in Fiji, argued that women should be central to development (Sen & Grown 1986). The value and importance of traditional knowledge and traditional social institutions are also becoming more recognised.

Fisheries development projects in the Pacific, like large-scale projects elsewhere in the world, have often had little impact, or worse, disastrous consequences. The frequent failure of rural development projects has been a topical issue in the Pacific Islands for years (Schoeffel 1985, Munro and Fakahau 1993, Veitayaki 2000). Debates rage on the appropriate role and nature of development projects, and the requirements for effective development. Whereas some would blame apparent failures on project design and process (e.g. inappropriate approaches, unrealistic goals) others point to the people in local communities (e.g. poverty, ethnic tensions, political inexperience). The general conclusion is that there has been a poor fit between the models of development and local aspirations and capacities.

In light of the ongoing argument, it is useful to ask whether or not the lack of major commercial success of fisheries development projects in the Pacific really amounts to "failure". The ice plant or wharf or local institution set in place by a project may cease to function when the funding ends but local people may nevertheless feel that they have benefited in less tangible ways. For example, access to new information, soft skills development, novel ideas that have impacts on indigenous institutions and relationships that can be employed by the people to advance their own priorities, are among the benefits that often go unrecognized. In addition, the lack of progress of a large-scale extractive fishery can be a blessing in disguise. This is particularly true when it prevents the overfishing of resources whose complex ecology is poorly understood.

From our case studies it is evident that, although development efforts may have been fitful and unrewarding from the point of view of developers, at the village level elders in particular perceive that change has been rapid and in some respects, poorly controlled or even regrettable. People value and seek out the various fruits of development such as improved health services, more rapid transportation, newly introduced fishing technologies, access to shops with modern consumer goods and schools for their children. However, benefits have not always been shared equally. Also, people recognize the costs incurred over time: the erosion of traditional values, loss of communal spirit, increased stress associated with the need for cash, and the ways in which fisheries resources have been overexploited to meet the new demands that accompany development. Modern media, which shows Pacific Islanders the apparently "rich and famous" of western societies, stimulates desires for goods and lifestyles that are difficult if not impossible to accommodate without putting at risk the ecology and cultural strengths of the Pacific (Emberson-Bain 1994).

Rosalie Masu is one of a new generation of fisheries officers in Solomon Islands. Here she is looking for fish to buy in Gizo, to take home to Honiara where prices are much higher.

Part of imagining a different and more effective or suitable approach to development in the Pacific is the recognition of the strength and opportunity embodied in Pacific cultures and vast marine territories (Waddell et al 1993). Traditional communities of the Pacific Islands have much to teach contemporary western societies, especially about self-sufficiency, sustainable living and community-based social supports. There is also much that Pacific Islanders can learn from one other. For example, Pacific communities have indigenous knowledge that can usefully augment modern science to provide the basis for effective resource management The recent success of the Locally Managed Marine Areas Network demonstrates the usefulness of the people's customary resource use arrangements in controlling resource use today.

FILLING THE RESEARCH GAPS

National planners and institutions have not afforded the small-scale fisheries much attention. These fishers are regarded only as villagers who fish when conditions compel them. Being part of the informal sector, small-scale fishers have no standing with lending institutions; it is argued that they lack the necessary securities, access to resources and communication skills. Thus, they lack the type of economic clout that could translate into political influence. The lack of attention to small-scale fishers is a situation that needs to change. Our case studies reveal that, even though many fishers are reliant on traditional, low-tech fishing practices and small, non motorized fishing canoes, fishing efforts over time have transformed inshore resources from undeveloped to fully exploited or depleted. The fishery has suffered from uncontrolled fishing effort, habitat destruction and land-based sources of pollution. The result is that food security is at risk on many islands, especially close to urban centres; this cannot be ignored.

Vendor selling fresh Caulerpa (nama) in Suva market, Fiji. This popular edible seaweed has been overfished to the point of local extinction in some Fijian villages, and so is a candidate for local resource management.

There is limited information on how development processes affect people and how the people have in turn shaped the development process. Research is required so that development practitioners, policy makers and managers can identify options that have a better fit with local culture, economic realities and expectations. The studies in this book are a contribution to such an effort; much more needs to be done.

INITIATIVES IN MANAGEMENT OF INSHORE FISHERIES

For development to be controlled and equitable there must be some form of resource management in place, preferably one that involves the people who are the targets of the development. There are many Pacific organizations now at work to transform the process of inshore fisheries development and management. These include USP, SPC, fisheries departments and non-governmental organizations (NGOs) such as the International Ocean

Institute's Pacific Islands Operational Centre and others that are involved in the Locally Managed Marine Areas Network.. With assistance from such outside agencies many communities, e.g. in Fiji, Samoa, Vanuatu and the Cook Islands, are resurrecting and reinventing community-based resource management institutions. In many countries, community leaders claim ownership of inshore fisheries resources and preside over resource use and management. Contemporary resource managers are seeking to mimic or to strengthen such traditional practices. It is therefore a particularly exciting and important time to be examining rural, small-scale fisheries of the region. The impact of new management interventions needs to be documented, so that Pacific Islanders can share their experiences and learn from the interesting transformations taking place around them. This publication is a step in achieving that.

PACIFIC VOICES

In the past, people have complained that most of what is known about the Pacific Islands has been written by people who are not indigenous to this region. Too often, Pacific Island researchers have allowed their work to remain locked up in academic theses or to languish in the "gray literature", inaccessible to the general public. In our project, writers have been supported not just to conduct research but also to publish their results. As project leaders, we firmly believe that Pacific Islanders need to do this more often: to select important topics, perform their own research and promote their work through formal publication.

SPECIAL THANKS

All the case studies in this book went through a process of review and editing. We owe a great debt to our reviewers for their wisdom, expertise and generosity. They were Jean Mitchell (UPEI), Elizabeth Matthews (Palau Conservation Society), Vina Ram-Bidesi (MSP, USP), Patricia Kailoa (USP, MSP), Ruben Sulu (MSP, USP) and Temakei Tebano (MSP, USP). Thanks also to Temakei Tebano for help with the scientific names of Kiribati fish and shellfish and to Tony Chamberlain of USP for providing some photos from his collection.

Thanks to SOPAC for providing maps to illustrate the book.

REFERENCES

Emberson-Bain, 'A. 1994 (ed). *Sustainable Development or Malignant Growth? Perspectives of Pacific Island Women.* Suva: Marama Publications.

Matthews, E. 1995 (ed). *Fishing for Answers. Women and Fisheries in the Pacific Islands.* Suva: Women and Fisheries Network. P.71-90.

Sen, G. & C. Grown 1987. *Development, Crises and Alternative Visions: Third World Women's Perspectives.* Monthly Review Press, New York. 116p.

Waddell, E., V. Naidu & E. Hau'ofa (eds) 1993. *A New Oceania, Rediscovering Our Sea of Islands.* Suva: SSED,USP.

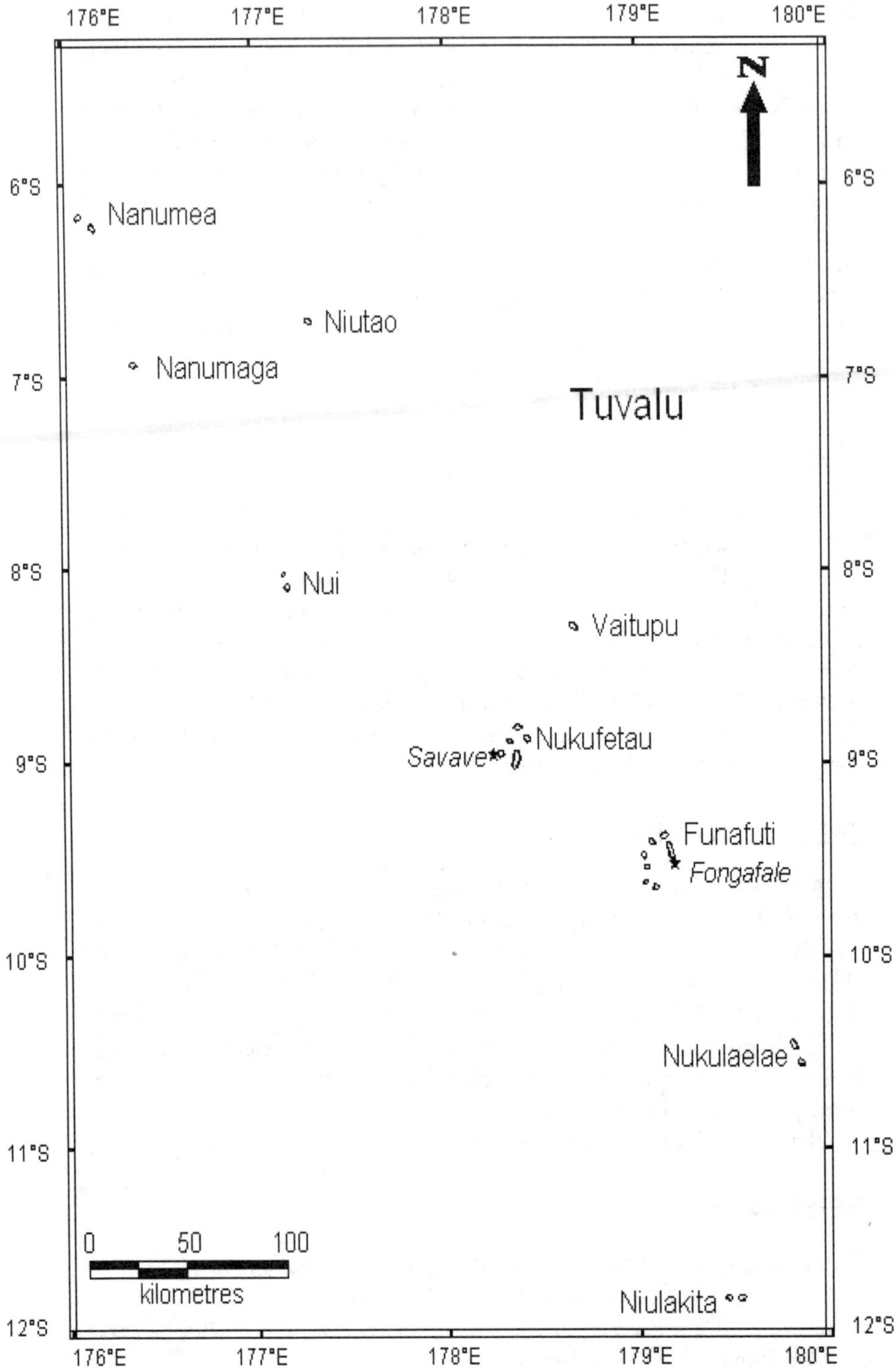

Map 1. Tuvalu, showing study sites (*)

2 Vulnerability And Dependence: The Nearshore Fisheries Of Tuvalu

Samasoni Sauni & Lilian Fay-Sauni

"... too much is going on in our seas and life is getting tougher by the day. I spend long hours out on the reef and at times on my canoe, only to discover that I can no longer ... fill my small woven basket with fish and invertebrates as I used to ... "

— *Nuausala Apelamo, Nukufetau fisherman, age 60*

THE FRAGILITY OF PACIFIC INSHORE FISHERIES

Awareness of how vulnerable Pacific reef and lagoon stocks are to overfishing has increased over the past decade. For instance, declines have been reported in reef and lagoon fish stocks in Palau (Johannes 1991), reef and lagoon fishes in Kiribati (Yeeting & Wright 1989), bonefishes, milkfishes and parrotfishes in the Cook Islands (Anon. 1988), various grouper stocks in Tokelau (Hooper 1985) and the Cook Islands (Sims et al 1990), and reef and pelagic fishes in Samoa (Helm 1992). Declining trends are reported in at least some of Tuvalu's reef fisheries (see quote above), and reductions in stocks of giant clams and bêche-de-mer have been recorded (Belhadjali 1998).

It seems that improvements in the fishing power of gear and growth of human populations in many locations have been paralleled by declines in stocks, catch rates and, in some cases, landed volume of reef fishes. What this means to coastal communities is not clear because there is little hard information on the degree to which people are dependent on marine resources. This lack of information is particularly the case for many small island states in the South Pacific such as Tuvalu.

People's dependence on marine resources may be measured in terms of the percentage of cash income that comes from fishing and by the degree to which fish is used to fulfil basic protein requirements. This study focuses on nearshore fisheries of Tuvalu. We investigate the degree of dependence on these fisheries for both food and income on two atolls, one urbanized and one rural. Results allow us to compare current catch and consumption rates with previously published data. We also consider our results in light of available information on the size of available standing stocks. This leads to a discussion of whether current fishing efforts are sustainable, and what management approaches may be required.

CASE STUDY RESEARCH METHODS

Fully structured household questionnaires were used to collect data on fish catch and consumption on Funafuti and Nukufetau atolls in July and August 1997. Sampling was haphazard and aimed to capture as many households as possible. Only those households that did not have any adult home at the time of the surveys were omitted. There was no bias toward fishing households and the survey covered people of both Tuvaluan and non-Tuvaluan origin. The 164 households included in the Funafuti survey represented 33% of the households on the island. On Nukufetau coverage was 50% (72 households).

In addition to household surveys, a creel survey was performed on all the handlining and gillnetting boats that landed at the sampling site on Funafuti over a 15 day period. A total of 45 boats were surveyed. These included boats owned by individuals as well as boats that were part of the commercial fishing fleet. The number of individuals on each boat, fishing time and gear used were recorded. Species, weight and number of fishes caught were recorded. Household data had showed that gill net and handline fishing were associated with large catches. The creel survey data were compared to catch and effort levels recorded in household surveys.

An independent consumption survey was carried out on a random selection of 100 households. Respondents recorded what they consumed in every meal over a period of one week. While it was not possible to issue every household with a set of scales to weigh the fish consumed, respondents recorded the average length of each fish using a graduated 0-15 cm scale drawn along the side of the survey form. Measured fish lengths were converted to estimated fish weight using a known length-weight relationship (Kulbicki et. al. 1993).

Microsoft Minitab was used to analyze rural and urban household data. Comparisons were made using Spearman Rank Correlation and Kruskal-Wallis tests. Fishing methods were assigned 'scores' depending on the importance of the method to the household, i.e. how often the method was employed. Thus, with seven fishing methods reported, a score of seven corresponded to the most important method; the score reduced by one for successively lower ranks. We calculated a weighted index of relative importance for each fishing method by summing the scores for each method across all households and dividing the total by the number of fishing households. Fishing households were those that used at least one fishing technique. An index was similarly calculated to compare the importance of various sources of household income.

Average fishing time, catch and fishing effort were calculated from both the household and creel surveys. Knowing the number of fishermen per boat trip and the fishing time, it was possible to estimate the mean catch per fishing boat and the mean catch per person-hour. Such estimates are based on the assumption that there was no prolonged adverse weather that affected fishing during the sampling period.

OVERVIEW OF CONDITIONS IN TUVALU

Geography and Economy

Tuvalu, formerly known as the Ellice Islands under the British Protectorate, consists of nine coral islands and atolls (Map 1). Total land area is a mere 26 km^2, but the country's Exclusive Economic Zone (EEZ) extends over an estimated 900,000 km^2 of the South Pacific Ocean (Anon. 1993). The islands of Tuvalu are scattered diagonally across 741 km of ocean, with distances of 40-100 km between any two adjacent islands. Most of the country lies less than three meters above sea level.

Tuvalu's shallow marine environment is dominated by fringing, barrier and intermediate patch reefs. Patch reefs and relatively barren coralline sand flats within shallow lagoon waters ($\leq$50 m) are surrounded by deep open ocean. There is virtually no continental shelf seaward of any of the islands. Water depths plunge rapidly to over 1,000 m within a few kilometers from the shore or outer reefs. The only substantial areas of shoal water are found in the internal lagoons.

The economy of Tuvalu is best characterized as traditional, non-cash and subsistence. There is a high dependence on public-sector employment and on remittances sent home by family members working overseas. In rural areas copra production, fishing and handicrafts may provide the only options for part time cash employment. People engage in subsistence activities such as fishing, agriculture and firewood gathering. The population of Tuvalu is around 9,561 (Government of Tuvalu, 2002), with an annual population growth of 0.5%. Over half of the population (53% or 4,493) now resides on Funafuti atoll, which has an annual population growth of 1.4 %. The remainder of the population resides on rural, outer islands and atolls. There has been a movement of people into the capital since the 1991 census.

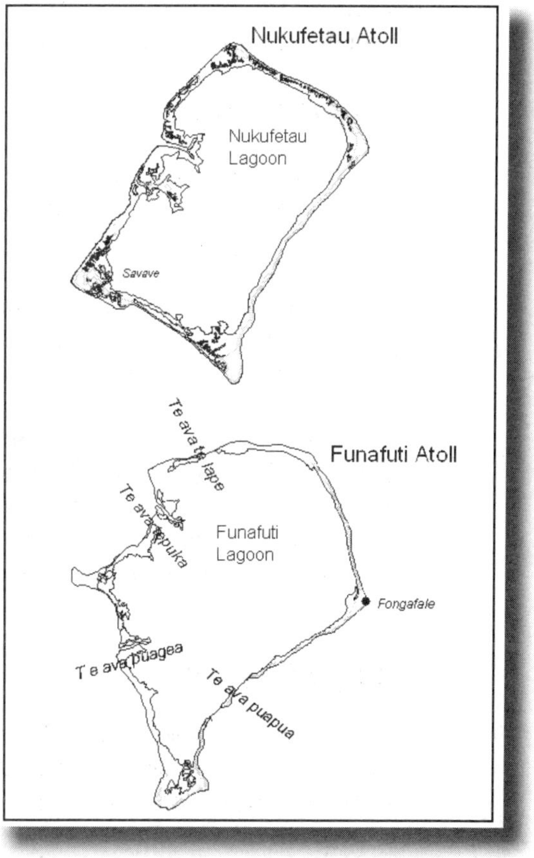

Map 2. Nukufetau and Funafuti Atolls lie next to one another in southeastern Tuvalu.

Tuvalu's Fisheries

Inshore fisheries of Tuvalu may be categorized as subsistence, artisanal or semi-commercial. Fishers use small canoes and boats that rarely reach ten meters in length. These crafts are propelled by paddles, oars, sails or outboard engines. The size of vessels limits fishing to inshore waters. Catches consist predominantly of pelagic species including varieties of tuna and flying fish that are fished using pole and line or trolling. Demersal species are caught by handlining on reef formations. Netting and spearfishing are restricted to shallow reef areas. Total consumption of inshore catches was estimated to be 730 tons in 1994 (Berdach & Maynard 1994). Offshore commercial fisheries in Tuvalu's EEZ are licensed and monitored. However, data on inshore fisheries are scarce.

Some options for mariculture and aquaculture have been tested, notably for seaweed, giant clams, trochus (topshell) and pearl oysters. The intention was to enhance productivity in the wild in response to reductions in available stocks, particularly of commercial species of bêche-de-mer, pearl oysters and giant clams (Belhadjali 1998).

Fisheries and related matters are governed under the Fisheries Act and its subsidiary provisions, which highlight the power of the government to protect fish and to promote, regulate and control fishing industries. In addition, Island Councils establish bylaws to regulate local fishing activities. These bylaws are perhaps crucial in strengthening local management of fisheries resources (Sauni 2000).

FUNAFUTI ATOLL

Funafuti, the capital, occupies the largest atoll in Tuvalu. The atoll has a land area of 2.7 km^2 divided among 33 separate islets (motu), the two largest of which (Fongafale and Tengako) account for 60% of the total land area (Map 2). The main human settlement is on Fongafale, where the investigation was undertaken. Households surveyed on Funafuti represented 1,227 people out of a total population of about 3,839 (Tesfagiorhis 1991). There were on average 7.7 persons per household, which is slightly lower than recorded in the 1991 census.

The reef platform on the eastern side of the atoll is quite narrow (300-400 m) and continuous; the islets are long and narrow. On the western side, the reef is wider and more broken and motu are more widely spaced, variable in shape and compact. The lagoon is by far the largest in Tuvalu and occupies an area of about 205 km^2 (Mclean & Hosking 1991).

NUKUFETAU ATOLL

Nukufetau, located roughly 120 km northwest of Funafuti, is an atoll consisting of 37 motu with a total land area of 3.3 km^2. Households surveyed on Funafuti represented 376 people out of a total population of about 751 (Tesfagiorhis 1991). A mean of 5.2 persons per household was recorded.

Nukufetau ranks second to Funafuti in terms of reef and lagoon areas (Map 2). Within the outer reef perimeter is an area of about 116.5 km^2, of which 22 % is reef platform and 78 %, the enclosed lagoon. The lagoon has a single surface channel (Te Ava Amua) from the open sea, but it also fills and drains across the reef flat and through subterranean passages. The bulk of the reef platform (85 %) consists of bare reef flat; vegetated islands and adjacent beaches account for the remainder (Mclean & Hosking 1991).

DEPENDENCE ON INSHORE FISHERIES

Economic Importance of Marine Product Sales

Only a small percentage of households on each atoll did not participate in any form of fishing. Participation in fishing, especially subsistence fishing, was high on both Funafuti and Nukufetau. This was true in terms of the numbers of households that fish (Table 1) and in terms of the frequency of fishing trips (Graph 1).

Table 1. Percentage of households engaged in subsistence and artisanal fishing on the atolls of Funafuti and Nukufetau

	FUNAFUTI (n=164)		NUKUFETAU (n=72)	
	H'holds	Frequency (%)	H'holds	Frequency (%)
No fishing	21	13	6	8
Subsistence	143	87	66	92
Subsistence + artisanal	23	14	48	67

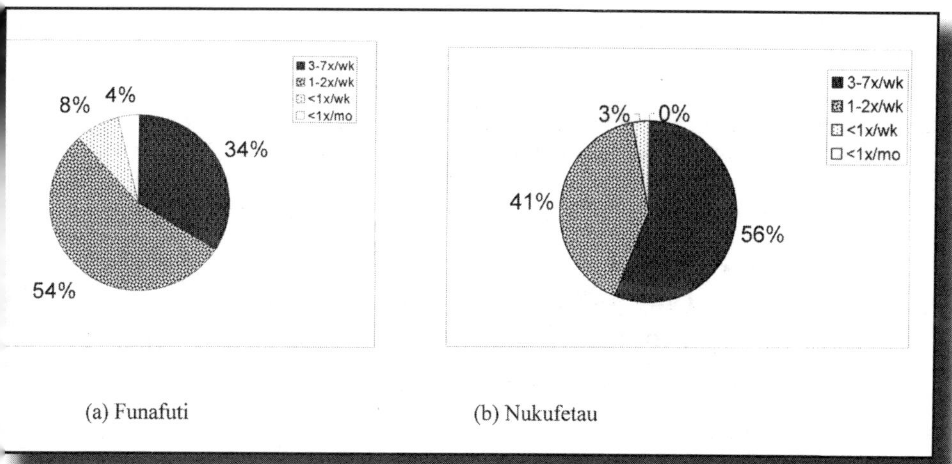

Graph 1. Frequency of participation in fishing on (a) Funafuti and (b) Nukufetau.

Whereas Funafuti households focused on subsistence fishing, a larger proportion of Nukufetau households engaged in artisanal as well as subsistence fishing. On Nukufetau, fishing was also more frequent. Whereas a majority of households on Funafuti fished once or twice a week (Graph 1a), the majority on Nukufetau fished three or more times per week (Graph 1b).

When all marine products were considered (i.e. processed as well as fresh fish and invertebrates) we found that the sale of marine products for household income was twice as common on rural Nukufetau compared to urban Funafuti (Table 2, Graph 2). Close to half of Nukufetau households regarded selling marine products as their most important source of income. A significant minority was entirely dependent on marine products as a source of income. This was a much greater degree of dependence on fishing for income than was seen in Funafuti.

Table 2. Income generating activities reported by households (HH) on Funafuti (Fun) and Nukufetau (Nuk).

		HOUSEHOLD ACTIVITY				
		Sale of marine products	Wage employment	Own business	Remittances*	Others
% Households involved	Fun	43.3	81.1	30.5	39.0	5.5
	Nuk	91.2	26.5	14.7	61.8	26.4
Most important source of income (% of HHs)	Fun	8.5	77.4	5.5	4.9	0.0
	Nuk	44.1	14.7	32.4	5.9	5.9
% Households depending solely on activity	Fun	3.0	17.7	1.8	2.4	0.0
	Nuk	14.7	-	-	-	-

* Money sent by family or friends working overseas

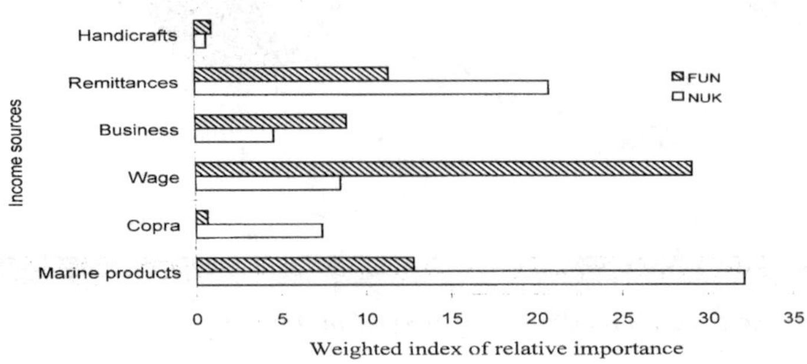

Graph 2. Relative importance of various sources of income to households on Funafuti and Nukufetau.

In addition to fishing income, many households on both islands received remittances from overseas. This is most prevalent on Nukufetau where many residents must move away to find paid employment (Resture & Resture 2004). On Funafuti, by contrast, wage employment was the most important source of income (Graph 2, Table 2). Other lesser sources of income included small business, copra production and handicrafts.

From among all exploited marine products, the sale of fresh fish was more prominent in Nukufetau than on Funafuti (Table 3). Fresh fish was the most important marine product sold by the majority of Nukufetau households (Table 3). Most fresh fish sold on Nukufetau goes directly to the Community Fish Center, which in turn transports the fish to the National Fisheries Corporation of Tuvalu on Funafuti.

Some households on both atolls processed marine products for sale. Gender division of labour was apparent in the case of these value-added products. Men harvested fish, shark fins and bêche-de-mer while women attended to drying and processing (pers obs). The harvesting of shells for handicraft production was also women's work. In our sample, roughly one third of households on Nukufetau and more than half on Funafuti generated some income from women's shell fishing activity (Table 3). For many of the Funafuti households, shells were deemed to be the most important marine product used to generate income.

Table 3. Households (HH) involved in trading specific marine products (year 1997).

	Fresh fish	Salt dried fish	Bêche de mer	Sharkfins	Ornamental shell
Funafuti					
% HH involved	26.7	11.6	1.2	2.3	58.2
Most important traded product (% of HHs)	30.1	13.7	1.4	0	54.8
Nukufetau					
% HH involved	45.0	13.3	3.3	1.7	36.7
Most important traded product (% of HHs)	78.8	6.1	0	0	15.1

Seafood Consumption

Fish was a key source of protein in both study sites. The majority of households on Nukufetau consumed marine products every day. Only one tenth of Nukufetau households restricted consumption to fewer than three times per week (Graph 3b). On Funafuti, one third of households ate fish every day while a third ate fish fewer than three times a week (Graph 3a). This difference in consumption patterns between the two atolls was not statistically significant (Kruskall-Wallis test: $p > 0.05$)..

Using consumption and meal survey data from Funafuti we estimated the per capita annual consumption of fish to be 2.9 kg $head^{-1}day^{-1}$. Assum-

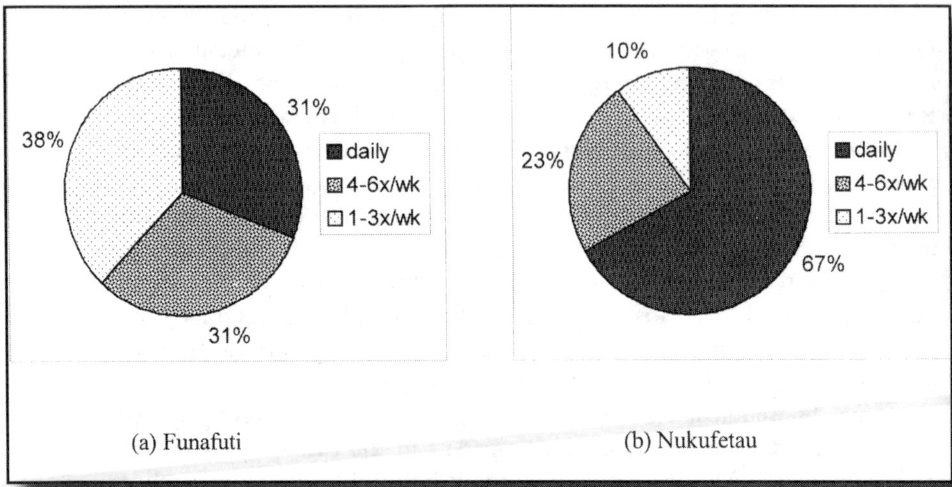

Graph 3. Frequency of marine product consumption at (a) Funafuti and (b) Nukufetau.

ing that fish consumption is constant through the year, extrapolation of these figures indicates a potential annual fish consumption of 4,052 t yr^{-1} for the Funafuti population alone (Sauni 1997). A second calculation, performed on a per household basis, provided an estimate of 0.9 kg $head^{-1}day^{-1}$ (Sauni 1997). Using this more conservative figure, annual consumption would be 1,258 t yr^{-1} for the population of Funafuti.

CATCH LEVELS AND FISHING EFFORT

According to household surveys, fish catch in July-August 1997 varied depending on which of the three commonly fished habitats was targeted: coastal reef, lagoon or open sea / outer reefs. On both islands, catches per trip from lagoons were relatively low, while catches from the open sea were higher and made up 11-15% of the overall catch (Table 4). Despite the lower catch per trip, people on both atolls, and especially on Nukufetau, fished most often in lagoons and least often in the open sea. The average weight of fish caught per trip on inshore reefs and for all habitats combined was significantly higher for Nukufetau than for Funafuti (Table 4).

The fishing technologies most commonly employed on both atolls were hand line (or drop line), gill net, trolling and fishing rod (Graph 4). Handlining refers to the use of a fishing line with a baited hook and sinker. Gillnetting is not a passive method in Tuvalu but involves a group of people driving fish (mainly schooling species) into the net. According to household surveys, handlining and gillnetting were predominant technologies, with handlining being more than twice as important as most other methods. Handlining and gillnetting can be carried out in all habitats of coastal, lagoon and outer reef fishing grounds.

The use of collection techniques involving bare hands or simple tools such as knives was prominent among women fishers. Survey respondents

Table 4. Estimated finfish catches per fishing site as reported in household surveys. Significant differences between Funafuti and Nukufetau (Kruskal-Wallis test) are also indicated (indicates p<0.05 ; NS = not significant).*

Habitat	FUNAFUTI Trips	Total Catch (kg)	Avg catch ($kg\ trip^{-1}$)	NUKUFETAU Trips	Total Catch (kg)	Avg Catch ($kg\ trip^{-1}$)	Significant Difference
Lagoon	139	4,519	32.5	68	2,310	34.0	NS
Reef	93	3,105	33.4	22	810	36.8	*
Open Sea	19	945	49.7	14	555	39.6	NS
Total	251	8,569	34.1	104	3,660	35.2	*

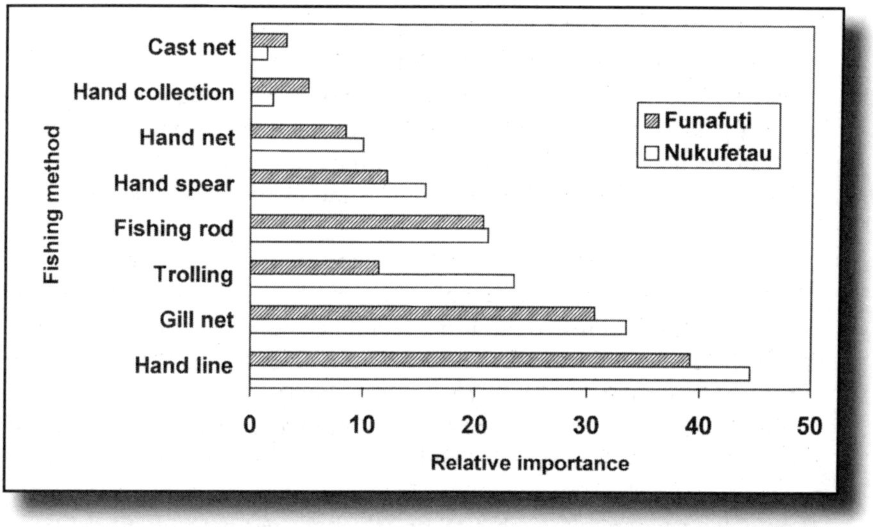

Graph 4. Weighted index of relative importance of fishing methods on two atolls, as determined from household surveys

on both atolls and particularly on rural Nukufetau, indicated that hand gathering was of relatively minor importance.

In the creel surveys, catch was recorded from handlining and gillnetting that was carried out on Funafuti's coastal reefs and lagoons. About 27 fish species were observed in the catch. Lagoon and reef species were prominent in terms of numbers of fish caught and weight (Table 5). Total catches were low for demersal species, large pelagics (average weight 1.6 kg) and small bait fish (average weight 0.3 kg).

It was not possible to establish catch per unit effort (CPUE) for each fishing gear and habitat because fishers at times used both gears and targeted more than one habitat in one fishing trip. However, fishers reported that their catch was predominantly gained through the use of handlines. Each fishing boat had, on average, 2.2 fishers.

Table 5. Summary of major groups of fish reported from the creel survey (Funafuti).

	Number of fish	Freq (%)	Weight (kg)	% of total weight	Average kg/ fish
Pelagics	153	3.0	151.1	7.7	1.6
Demersal species	383	7.5	337.3	17.3	1.0
Lagoon & reef fish	4,096	80.1	1,304.9	66.9	0.7
Bait fish	484	9.5	156.8	8.0	0.3
Total	5,116	100	1,950.1	100	

Creel surveys on Funafuti recorded an average catch of 43.3 kg per fishing trip, which is about 25 % higher than that reported for reef and lagoon fishing trips in the household surveys. Creel data are more reliable as they are actual weight measurements whereas household data are estimates provided by people who may not commonly use a weighing scale. It is however possible that the household data are reasonably accurate. Household catches may be lower because the fishing gears in use were more varied and possibly less efficient than the gillnets and handlines whose catch was reflected in creel surveys.

A total catch of 1,950 kg was recorded for 98 individuals taking part in 45 fishing trips which averaged 9 hours in length (Table 6). For his efforts, the average Funafuti fisherman using handline and gillnet took home 19.7 kg of fish per day and was able to extract on average 2.2 kg of fish per hour of fishing (equivalent to 4.8 kg per hour per boat).

Respondents to household surveys estimated that the length of their fishing trips averaged 5.3 and 7.9 hr on Funafuti and Nukufetau respectively. Putting this together with the average catch per trip (Table 4) provides estimates of catch rate, or kg caught per hour of effort (Table 6). The Funafuti figure for catch rate is higher than for Nukufetau. Even

Table 6. Estimates of fishing effort and catch rate from creel and household surveys.

Survey	Total # Fishers	Total Time (hr)	Total # Trips	Effort (avg hr per trip)	Catch rate per trip (kg/hr/boat)	# fishers per trip	Catch rate per fisher (kg/hr/person)
Creel Survey							
Funafuti	98	404	45	9	4.8	2.2	2.2
Household Surveys							
Funafuti	316	759	251	5.3	6.4	1.3	4.9
Nukufetau	136	411	104	7.9	4.5	1.3	3.5

though the average catch per trip is lower on Funafuti, trip length is much shorter.

Catch rate is greater when calculated from the Funafuti household survey than from the Funafuti creel survey. Creel survey catches are higher but this is offset by the fact that fishing trips are longer and involve more people. As noted above, differences between creel and household data may also be ascribed to different levels of accuracy and/or to the different mix of fishing gears being documented.

DISCUSSION

Fishing Activities in Rural and Urban Atolls

In Tuvaluan households, it is not common to earn a living from trading only one product. Household members may be formally employed, work on a plantation, or fish but most commonly perform a combination of roles.

A wide array of gears were used to catch fish, but hook and line, followed by gill nets, were most commonly used on both atolls. In both Funafuti and Nukufetau, people preferred to fish in lagoons and coastal reefs, despite lower catch rates in these inshore habitats relative to the open sea and deep reefs. Reef and lagoon species also predominate in creel surveys of Funafuti fishers. This is perhaps because costs (fuel and time) are higher for deep water fishing. It may also signal that some demersal, pelagic and bait fishes are less abundant, either in general or in the season sampled. Alternatively, it could reflect the fishers' preference for eating or selling inshore species, or a seasonal preference for fishing in particular habitats. Aswani (1998) described how fishers in Solomon Islands switched fishing grounds according to season, with the May through September period being the peak time for targeting inshore reefs and lagoons because of the occurrence of low tides during the day. A longer term and more in-depth study would be required to fully understand what is going on in Tuvalu.

Catches per trip on Funafuti's open sea (i.e. trolling for pelagic species and handlining for deepwater snappers and related bottom dwellers) were relatively high but variable and therefore not, on average, significantly different from Nukufetau. By contrast, catches per fishing trip targeting inshore reefs were significantly higher on Nukufetau. Average catch per hour from all fisheries was, however, higher in Funafuti. This is at first sight puzzling but in Funafuti, households reported some very high catches (averaging almost 50 kg per trip) from open sea fishing and these will have pushed up the overall catch per hour figures. Also, in Nukufetau where fishers have little else to do but fish they may stay at sea even when fishing is not very productive. In Funafuti, by contrast, fishers often have day jobs and therefore have limited time for fishing and must be efficient. Funafuti households reported relatively shorter fishing trips. Commercial boats sampled in the creel survey fished for longer periods, employed more fishers per trip and landed larger catches per trip.

Gender Roles in the Fishery

The study shows how both men and women have economic roles in inshore fisheries. While men are primarily fishing in reef and lagoon areas using a variety of technologies, the women collect inshore food species with their bare hands, sometimes aided by simple tools. Women are active in processing fish, bêche de mer and other species and they also collect and process ornamental shells for income. Women's fishing contributes an income source to many households, especially in urban Funafuti where ornamental shells and handicrafts are frequently the most important income-generating marine products. The high proportion of households active in the ornamental shell trade in Funafuti can be explained by the fact that the bulk of shells traded come in from the outer islands to the capital, which is the main market (see also Resture & Resture 2004). The income generated from shell handicrafts is low compared to income from the sale of fresh fish. Also, harvesters can gather 1,000 shells in just one hour, so fishing time per week is much less than for food fish (Resture & Resture 2004). This helps to explain why women's hand gathering is deemed a relatively unimportant fishing method.

Dependence on Marine Resources in Tuvalu

Fish and marine products made up an important component of the diet of Tuvaluans surveyed. Households on Funafuti consumed fish at least once a week while a majority of Nukufetau households consumed fish every day. The higher rates of fish consumption on Nukufetau can be explained by the limited alternative sources of protein and the very frequent fishing activity. This may apply equally to other outer islands and atolls. Alternative sources of protein, primarily imported, were found on Funafuti. These would logically replace some local fish consumption on this island.

The high dependence on fish on both atolls may have different root causes. Dependence on fisheries for income was particularly apparent on rural Nukufetau, where reef resources are more abundant and alternative means of earning cash are few. After world copra prices collapsed there was a shift from copra production to fishing as a source of rural incomes. This helps explain the currently high dependence on fisheries. Rural fishing is thus critically important both as a source of income and as a source of protein.

On urban Funafuti the increasing population, fed by both high birth rate and in-migration, may be the key factor driving an increasing demand for fish. The ability to switch to imported meat or canned fish is dependent on each household's cash earnings. Given the limited options for generating cash income it is unlikely that a family could depend entirely on imported meat. A high percentage of urban households still catch their own fish to eat but relatively few have surplus to sell. Many people on Funafuti are in jobs during the day and go fishing at night and on weekends. They fish not only close to home but also out at sea. As long as there is demand for pelagic species, households with trolling gears can be expected to participate in this fishery, even though it occupies many hours and requires searching out schools of fish at a great distance from land.

Changes in Exploitation and Consumption over Time

A key finding of this study is that fish catch rates have increased if compared to past estimates. For instance, weight of catch per trip reported from the creel survey was much higher (by 64 %) than rates extrapolated from annual catch estimates for 1993 and 1994 (Dalzell et al 1996). Similarly, fishing assets per household and use of motorized, aluminum or plywood boats were greater compared to the 1980s and early 1990s.

Our estimates of the weight of fish consumed per capita (0.9 – 2.9 kg per day) on Funafuti atoll appear to be 44 - 83 % greater than reported in 1993, when per capita consumption of fish on Funafuti was estimated at 0.5 kg per day (Anon. 1993). Local fisheries officers using questionnaire surveys collected the 1993 data from Funafuti. High rates of fish consumption in Tuvalu could be driven by high catch rates but it is not known whether increased catches signify healthy stocks or increased fishing efforts that might endanger stocks.

Levels of fish consumption are high but variable across the Pacific. Zann (1983) estimated the average per capita consumption of whole fish at 0.6 kg per day on rural atolls of Kiribati and 0.3 kg per day in urban South Tarawa of Kiribati. Rawlinson et al (1992) reported an average per capita consumption of 0.2 kg per day for Fiji.

Vulnerability of Marine Resources

The sustainability of fisheries in Tuvalu is difficult to assess because the few catch and stock estimates vary considerably. Unexploited demersal stocks in Funafuti were estimated at 224 tons (Dalzell & Preston 1992). All fishable habitats and seamounts in Tuvalu were estimated to be capable of providing 76 t yr^{-1}; a fishable stock of 8 t yr^{-1} was estimated for Funafuti alone (King 1995).

In 1993 the estimated annual per capita consumption of fish was 182.5 kg, signifying total consumption of around 730 t yr^{-1} (Berdach & Maynard 1994). Reported catches in the artisanal sector at that time were only 300 t yr^{-1}. Berdach & Maynard (1994) believed that the balance of 430 t yr^{-1} was being supplied through unreported subsistence fishing. Our estimates of fish consumption in 1997 imply an annual fish consumption of more than 1,258 t yr^{-1} for the Funafuti population alone (Sauni 1997). This exceeds all previous estimates of exploitable fish stocks, suggesting that rates of extraction are unsustainable.

THE NEED FOR FURTHER RESEARCH AND MANAGEMENT EFFORTS

The case study confirms the high dependence of both urban and rural Tuvaluans on inshore resources for food, and the particularly heavy dependence of rural Tuvaluans on fish for income. The findings also point to a dangerous level of vulnerability as stocks near the most densely populated atoll of the country, Funafuti, appear to be being fished at unsustainable rates. Pauly (1990) refers to "Malthusian overfishing" where available reef and lagoon fish cannot sustain a human population when it grows rapidly.

The introduction of modern technologies has not only increased pressure on existing resources but has also allowed access to previously inaccessible resources. This trend is in line with the depletion of inshore resources occurring widely in the Pacific region (Adams & Wata 1995).

Tuvaluans have little choice but to rely on inshore fisheries as the main source of protein for the foreseeable future. However, this will be difficult, as there is likely to be an increasing volume of high-value species being transported to domestic urban and tourist centers and exported to overseas markets. The greatest problems are likely to occur on Funafuti where yields from inshore areas may not keep pace with population or tourism growth. To date there have been no major efforts to target offshore fish stocks, or to develop aquaculture that might reduce pressure on traditional fisheries.

Unless some effective management measures are put in place, overfishing in inshore areas will continue to increase and could seriously reduce stocks of targeted species, ultimately jeopardising the future livelihoods of Tuvaluans.

Based on the results of this study we recommend the following:

- Independent assessment of the current status of target reef and lagoon species using underwater visual census,
- Independent assessment of catches from predominant fishing gears,
- Independent assessment of current status of ornamental shells in all atolls and islands of Tuvalu,
- Enforcement and policing of the Funafuti Conservation Area; establishment of a similar conservation model in Nukufetau and other atolls and islands with expressed interests, and
- A nearshore resource management plan prepared for each of the atolls and islands, based on available research information.
- It is further recommended that the management plans should be enacted under both the Falekaupule Act and the Fisheries Act and implemented locally by island and atoll communities, with technical assistance from government departments.

NOTES

1 defined as the sale of fresh and/or processed marine products including fishes and invertebrates. Not all households participated in sale of marine products; thus, fishing activities cannot be linked to the actual sale of marine products.

2 National Fisheries Corporation of Tuvalu

REFERENCES

Adams, T. & Wata 1995. Review of the FFA/ICOD Research Coordination Unit. In Research on Fisheries in the Pacific Islands Region. New Caledonia: South Pacific Commission. P.4-6.

Anonymous 1988. Report of the Workshop on Pacific Inshore Fishery Resources. South Pacific Commission, Noumea.

Anonymous 1993. Tuvalu Bottomfish Project: Biological Assessment Summary Year One. RDA International Consultancy Report. Placerville, USA.

Aswani, S. 1998. Patterns of marine harvest effort in southwestern New Georgia, Solomon Islands: resource management or optimal foraging? Ocean & Coastal Management 40:207-235.

Belhadjali, K. 1998. A Survey of the Inshore Fisheries Resources of Tuvalu. Unpub. report prepared for the Ministry of Natural Resources and Environment, Tuvalu. 101p.

Berdach, J.T. & J.A. Maynard. 1994. A Study of the Fisheries Sector of Tuvalu. Asian Development Bank (ADB) & Government of Tuvalu. P.1-9.

Dalzell, P. & G.L. Preston 1992. Deep Slope Fishery Resources of the South Pacific. Inshore Fisheries Research Technical Document No. 2. New Caledonia: SPC. P.233-250.

Dalzell, P., T. J. H. Adams & N. V. C. Polunin 1996. Coastal fisheries of the South Pacific Islands. In R.N. Ansell, Gibson & M. Barnes, Oceanography and Marine Biology: An Annual Review. London: UCL Press. 34: 395-531.

Government of Tuvalu unpublished. Census, 2002.

Helm, N. 1992. A report on the market survey of reef and lagoon fish catches in Western Samoa. In P. Dalzell (ed) Papers on Fisheries Science from the Pacific Islands. New Caledonia: South Pacific Commission. P.1-5.

Hooper, A. 1985. Tokelau fishing in traditional and modern contexts. In K. Ruddle & R. E. Johannes The Traditional Knowledge and Management of Coastal Systems in Asia and the Pacific. Jakarta: UNESCO Regional Office for Science & Technology for SE Asia. P.9-38.

Johannes, R. E. 1991. Some Suggested Management Initiatives in Palau's Nearshore Fisheries, and the Relevance of Traditional Management. Inshore Research Project unpublished consultancy report. New Caledonia: South Pacific Commission.

King, M. 1995. Management Plan for the Deepwater Bottomfish Fishery in Tuvalu. Unpub. discussion paper, Government of Tuvalu. 18p.

Kulbicki, M., G. Mou Tham, P. Thollot & L. Wantiez 1993. Length-weight relationships of fish from the lagoon of New Caledonia. Naga, 16(2-3):26-30.

Mclean R.F. & P.L. Hosking 1991. Geomorphology reef islands and atoll motu in Tuvalu. South Pacific Journal of Natural Science 11:167-189.

Pauly, D. 1990. On Malthusian overfishing. Naga 13(1):3-4.

Rawlinson, N. J. F., D. A. Milton., S. J. M. Blaber, A. Sesewa & S. P. Sharma 1992. A Survey of the Subsistence and Artisanal Fisheries in Rural Areas of Viti Levu, Fiji. ACIAR Monograph No. 35, 138p.

Resture A. & S. Resture 2004. Seashells on the seashore: Women's participation in the shell trade on Funafuti and Nukufetau, Tuvalu. In I. Novaczek, J. Mitchell & J. Veitayaki (eds) Pacific Voices: Equity and Sustainability in Pacific Islands Fisheries. Suva: IPS, USP.

Sauni, S. 1997. An Assessment of the Inshore Fisheries in Tuvalu with Reference to Funafuti and Nukufetau Atolls. MSc Thesis, University of Newcastle Upon Tyne, England.

Sauni, S. 2000. The status of the coral reefs of Tuvalu. In Coral Reefs in the Pacific: Status and Monitoring, Resources and Management. Proc. of a Regional Symposium International Coral Reef Initiative (ICRI) 22-24 May, 2000. New Caledonia: Noumea IRD Centre. P.331-349.

Sims, N.A., G.L. Preston & R. Alfred 1990. Pearl Oysters in Christmas Islands, Kiribati, and the Potential for Development of a Pearl Culture Industry. Inshore Fisheries Research Project Technical Document Draft. New Caledonia: SPC. 14p.

Tesfagiorhis, H. 1991. Tuvalu: 1991 Population and Housing Census. Government of Tuvalu 2: 1-5.

Yeeting, B & A. Wright 1989. A Review of the Status of Fisheries Resources in Tarawa Lagoon, Republic of Kiribati. FFA Report No. 89/16. Honiara: Forum Fisheries Agency.

Zann, L.P. 1983. Traditional Management and Conservation of Fisheries in Kiribati and Tuvalu atolls. Suva: Institute of Marine Resources USP. P.1-23.

3 No Shells, No Langalanga: Hard Times In Malaita, Solomon Islands

Kristina Fidali-Hickie and Cherie Whippy-Morris

"The shell money production defines the history, culture and future of the people of Langalanga. It is the main source of sustenance for the villagers. No more shells, no existence for the people of Langalanga"

— Hellen Galona, Lalana women's group, Langalanga Lagoon

SHELL MONEY IN SOLOMON ISLANDS

Historically, shells have been highly valued throughout Melanesia, both as ornaments and as a form of exchange used in culturally important dealings. When shells were used as ornaments for different parts of the body, the type of shells worn reflected one's clan affiliation. Shells used as a form of exchange were known as "shell money" and many different kinds have been in circulation. Most were made of small disks arranged on strings. The strings had different values depending on the type and colour of shell used in making the disks (Lewis, 1929). Shell money was used to barter for traditional foods such as yams, sweet potatoes, coconuts and pigs. It was also used to pay bride price in marriage ceremonies and as a payment to help settle quarrels and disputes. Such money was circulated in Papua New Guinea, Solomon Islands, northern Vanuatu and New Caledonia (Lewis, 1929).

In Solomon Islands (Map1), shell money was used extensively in the Shortland Islands and Malaita. Much exchange occurred between Malaita and other islands in the Solomons group and beyond them to Bougainville in Papua New Guinea. Women living on the shores of Langalanga Lagoon in Malaita engaged intensively in the production of shell money for trade. According to Connell (1977) the major driving force behind shell money production in Malaita in the 1970s was the demand from Bougainville. Cooper (1971) and Connell (1977) both suggested that shell money production was useful in that it bridged the gap between home-based, low-capital, traditional production and the cash economy. In addition, there were few economic alternatives for Malaitans in Langalanga.

Between 1965 and 1975 the value of shell money increased and production was facilitated by the introduction of modern tools. Despite the

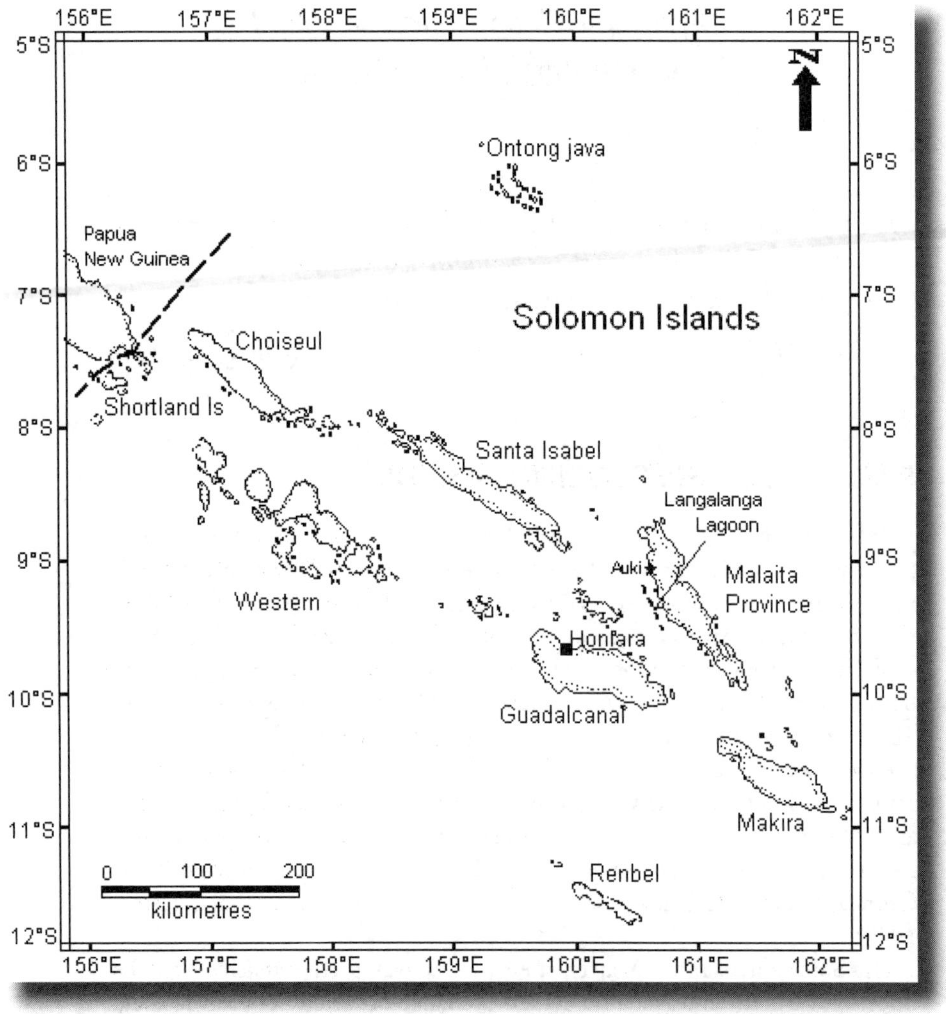

Map 1. Solomon Islands showing location of Langalanga Lagoon on Malaita.

increasing difficulty and expense of getting shells, more Malaitans from Langalanga were drawn into shell money production. Consequences of this increased production, as outlined by Connell (1977), included shortages in two types of shells (white and red) and a change in gender roles in the manufacturing process.

People in the Langalanga community have indicated that the shell money industry has changed in recent years. Some changes may be related to the social unrest in Solomon Islands that began in 1999 and resulted in a coup in 2000 and a change of government in 2001. These events have drained the country's economy and it is expected there will be repercussions for years to come. In such troubled times people desperate to meet basic needs often overexploit their natural resources. The breakdown of cultural traditions as a result of the introduction of new values and practices into the communities also contributed to changes in the shell money industry. Where traditional community resource management practices are absent or neglected and pressure to harvest resources is mounting, marine resource stocks such as the shellfish used in making shell money may be driven to the verge of local extinction.

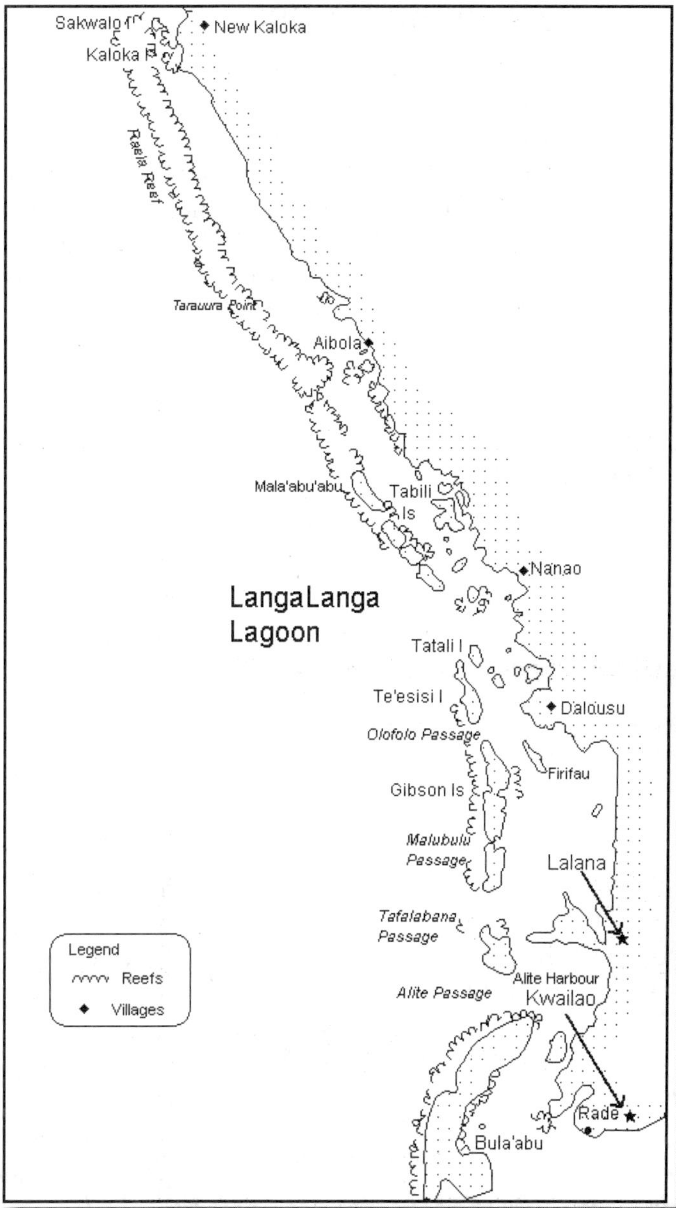

Map 2. Langalanga Lagoon study area.

This case study will look at the current conditions as well as the history of change in two villages, Lalana and Kwailao, where shell money is produced. Both villages lie within the Langalanga Lagoon area, which is the centre of shell money production in Malaita (Map 2).

CASE STUDY METHODS

Collection of data was conducted in October 2003, by direct observation, random household interviews and focus group discussions that involved various village leaders (e.g. elders, women's organizations, teachers, church ministers). In Kwailao, 27 of the 42 households were surveyed; in Lalana, 26 of the 79 households. In addition to providing basic demographic information, respondents were asked to describe their living conditions; their activities in the shell money industry, including harvesting, processing and marketing; their other sources of income and food; and their concerns and ideas for management of shell resources.

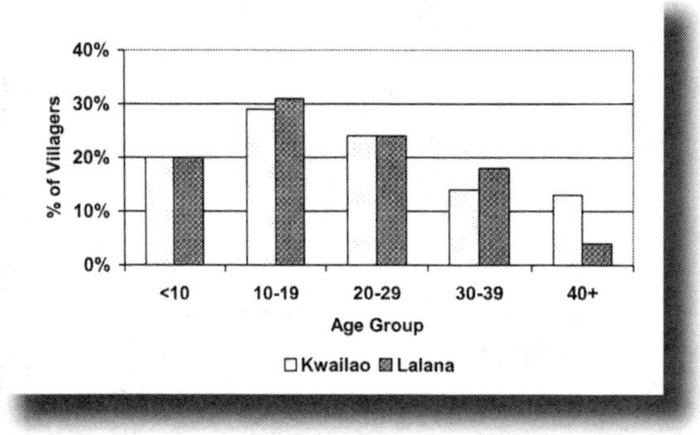

Graph 1. Percentage of villagers in surveyed households that were in each age category.

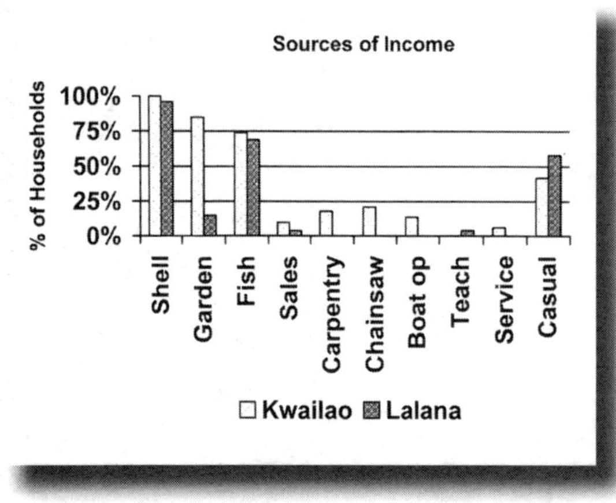

Graph 2. The percentage of surveyed households that had the following sources of income: shell money making, sale of garden produce, sale of fish, other small goods sales, carpentry/boat building, chainsaw operation, working on a boat, teaching, service jobs (e.g. house maid) and casual labour (stevedoring).

KWAILAO VILLAGE

Kwailao village is situated on the shore of Langalanga lagoon, facing the man-made island of Bula'abu, which is towards the southern end of a chain of islets (Map 2). The village population is divided among discrete hamlets, which grew in number from three to nine in the years between 2000 and 2003. This population increase was a result of migration from overcrowded Bula'abu Island and from other neighbouring villages, coupled with the return of Malaitans from Guadalcanal to their home villages during the period of social unrest, 2000 to 2002. From the household survey it was determined that almost half of the population is under 20 years of age and just over 10% is above 40 (Graph 1). Because the village has a mixture of people originating from different villages, there is no paramount chief. Notable village elders, church leaders and organized groups in the village perform the role of village organizers.

There is a rudimentary road linking Kwailao to other villages but people prefer to travel in outrigger canoes or motorized boats. Water for drinking, washing and laundry is mostly drawn from wells dug next to the houses, but some households have aluminium tanks to hold rainwater collected from roofs. Sanitation infrastructure is lacking.

The daily activities of the people of Kwailao revolve around shell money making, fishing (particularly for the men) and gardening. Everybody in the village who is over 10 years of age engages in shell money making as his/her major means of income generation. Although the villagers engage actively in gardening, most of the produce is for home consumption. Only if there is a surplus can the produce be exchanged for other food items or sold for cash. These and a few minor activities (Graph 2) are the basic economic options for the community.

The people depend heavily on the marine environment for their daily sustenance. Mangrove fruits are used to prepare a typical Langalanga soup, which may be eaten once a week. Protein is derived from fish and shellfish collected from the reefs and lagoon, while carbohydrate comes from root crops or is purchased from the market. In general, villagers eat only one meal a day. To eat three times a day would cost them too much money and time. A typical meal consists of potatoes or swampy taro baked in an earth oven, or a pot of rice taken with tea or mangrove fruits. On very rare occasions they will eat fish. For the most part, the fish they catch is sold. Because of their poor diet the people are not healthy; generally they appear frail, not vibrant. The people are prone to many diseases and malaria is common.

LALANA VILLAGE

It takes less than 10 minutes by motor boat to travel from Kwailao north to the village of Lalana. Unlike Kwailao, Lalana village is made up entirely of one group of people. They migrated in from Laulasi Island because of overcrowding on that island. When they arrived they used stones to build extensions to the shore, and settled on top of them. Lalana is led by a chief, village elders and church leaders. As in Kwailao, half of the population is under the age of 20 (Graph 1), but Lalana has fewer elders (<5%).

Lalana is renowned as a place where the shell money trade is "like a religion" to the people. Without exception, every household has highly skilled shell money producers. People reported that they spend 95% of their time making shell money. Because of the very limited land and the swampy nature of the area most villagers do not own a garden. For them to be able to move further inland to fertile soil, they would need to get permission from the inland landowners, who are Kwaraae speaking people, so they are restricted to their coastal dwelling places, which leaves them with few or no economic alternatives to pursue (Graph 2). This situation has led them to focus intensely on shell money making to generate cash income.

Lalana village is surrounded by mangrove trees, which provide fruits for making soup and firewood for cooking fuel. When men go fishing, their catch is sold and the money is used to buy food from the shops. The people depend on store-bought, refined food for the bulk of their diet. As in Kwailao, meals consist mostly of carbohydrate. The people generally lack proper sanitation and the water in their wells is brackish. Getting access to sufficient clean water is a persistent problem. People ration the use of water for drinking, cooking and other uses. Poor diet, poor water and swampy surroundings that breed malaria mosquitoes combine to undermine the health of the people.

BARRIERS TO SMALL ENTERPRISE DEVELOPMENT

Both villages face the same problem of limited resources for alternative livelihoods. In addition, the people say that their cultural (*wantok*) system makes it difficult to succeed in developing any small business. Some villagers have tried to start small enterprises - for example, small canteens, a bakery or cigarette selling - but most found that the businesses do not work in the long run. The problem is that *wantoks*, the members of one's extended family, take goods on credit and promise to repay when they finally sell their shell money. The borrowers accumulate so many debts that they are unable to repay them all when they finally have money. This chronic lack of cash flow seldom fails to shut down peoples' small enterprises. Carpentry is a notable exception. Shell money can be used directly to buy building materials, so villagers clear the mangroves from islets to make room for small boat-building yards. People from the Langalanga area are known for their boat building skills.

Faustina Kei holds the traditional form of shell money called akwala afu or tafuliae.

ATTACHMENT TO THE SHELL MONEY INDUSTRY

Kwailao and Lalana villagers are very active shell money traders. The women in particular feel strongly about the survival of the trade as their major source of income. In spite of a growing problem of resource scarcity and, worse still, the local extinction of some shells, they are determined to keep the trade alive even to the point where there are no more shells left in the sea.

Faced with meeting basic family needs as well as social and religious obligations, the women do not have much choice. Culturally, women are expected to ensure that there is food on the table and a bed for family members to sleep on and they are under pressure to live up to this expectation. As a result they exert pressure on the shell resource, which is seen as their only workable option for earning income. As will be seen below, the already massive pressure on the shell resource is made worse by the collapse of the traditional resource management system that historically helped to conserve shells.

SHELL MONEY PRODUCTION

Harvesting

Four different types of shell provide the coloured disks used in the making of shell money. A red-lipped rock oyster called romu *(Chama pacifica)* provides red coloured disks. Applying heat to a small white shell known locally as kee *(Beguina semi-orbiculata)* makes orange disks. Black disks are taken from large horse mussel shells called kurila *(Atrina vexillum)*. Thick white disks are obtained from a ridged cockle known as kakadu *(Anadara granosa)*.

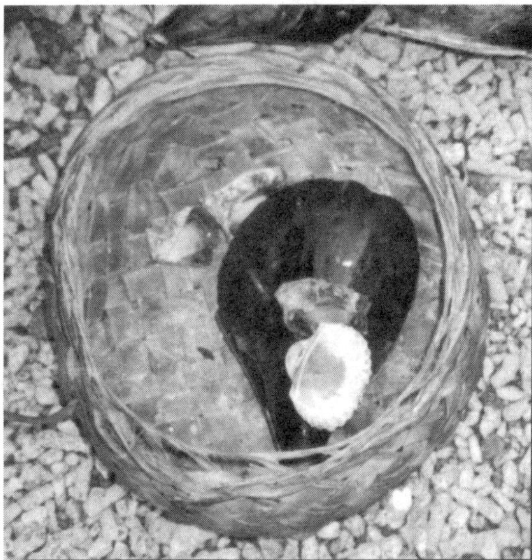

Shells used in the production of shell money : romu (red lipped oyster), kurila (large black mussel) and kakadu (ridged white cockle).

The four different shells inhabit different habitats in the sea. The kakadu, kurila and kee grow in shallow lagoon waters. Most are hand collected but kee grows attached by byssal threads to hard surfaces and therefore must be chipped off using a knife or screwdriver. The romu is found in deeper water towards the edge of the reef slope. Romu also grow attached to hard substrates and are collected using a chisel, screwdriver or metal bar.

All the shells are edible and there are other types of shells, most commonly giant clams, which are collected for food when diving. Some people also report that they pick up sponges to use in the kitchen when they dive for shells.

Kee shells which are heated to provide orange disks.

Cracking shells into chunks using an iron rod. In the past women used a palm-sized rock like the one lying in front of the large cracking stone.

Women chipping chunks of shell into round shapes using an iron rod.

Faustina Kei displays a traditional wooden drill while Sara Okofala employs a modern drill to pierce the shell disks.

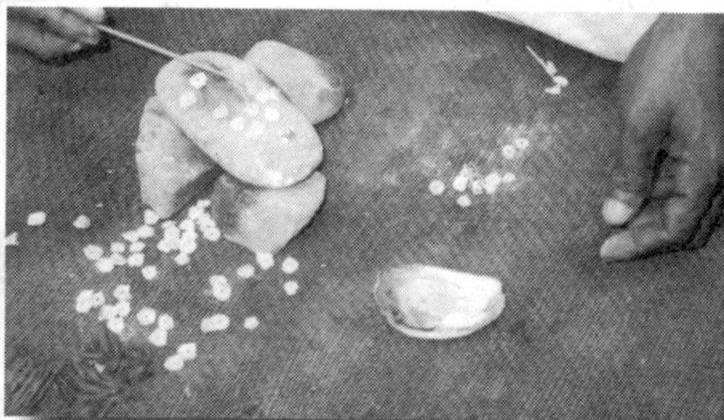

Applying heat to kee disks to turn them orange.

Smoothing kakadu disks by rubbing them between flat stones.

The Chinese stone used for sharpening.

Angel Kwaefala sharpening a string of disks.

Processing

Shell money is the end product of a long, labour intensive process that has many steps. The process is begun by cracking the four different types of shells into chunks. This is done carefully to minimize waste. Only particular parts of each shell are retained. After cracking, the bigger chunks of shell are chipped into smaller pieces, carefully shaped into round discs using an iron rod and drilled through the centre. The disks made of *kee* shells are heated over a hot flat stone to give the desired orange-brown colour. The deeply ridged *Kakadu* shell disks are flattened by rubbing them between a small stone and a large grinding stone. After this, each type of shell is separately threaded onto a nylon string. The threading of the shells for sharpening is good fun for the children, although adults also do this task. The string is tied against a piece of timber to be ground down, or sharpened, to a uniform size by rubbing with a "Chinese stone". The variously coloured shell disks are now ready to be threaded into shell money by a skilled man or woman.

Final Production of Shell Money and Ornaments

Traditional Langalanga shell money, *akwala afu*, which is also commonly called *tafuliae* by the Lau and Kwaraae people, is made up of ten strings of red, black, white and orange coloured shell disks threaded together in particular patterns. These strands of shell money usually range in length from two to three metres, but may occasionally reach four metres in length. People prefer to make the shorter strands to conserve on raw materials. Another traditional product is the *safi*, which is used by people in South Malaita and Guadalcanal. This is a single strand about two metres long that is composed only of *kee* shells and sells for SI$50 (US$6.75)[1].

If disks are not directly used for making shell money they may be sold to other people to make any number of different items. A string of 30 disks is sold for SI$0.30 - 0.40, the price being dependent on the colour of the shell. People in Langalanga make a variety of jewellery and ornaments from the shell disks. Some are for traditional purposes such as decorating a bride or celebrating special occasions. Priestly vestments may also be made of shell money. Other shell items including necklaces, earrings and bracelets are sold in tourist market places.

Shell ornaments worn by members of a pipe band welcoming visitors to the airport in Honiara, Solomon Islands.

CHANGING GENDER ROLES IN THE SHELL MONEY INDUSTRY

Historically, women and men had specific roles in shell money production. These roles were based on the cultural expectations of the community. The role of men was to go to sea to dive for shells, either near their village or on more distant reefs. In modern times, women have become active in diving and collecting shells and may even travel to distant islands to engage in harvesting.

In the past, when the men arrived home, women would dry the shells and put them away. The women then performed the tasks of cracking the shells, shaping, drilling, heating and threading, while men might help with smoothing and sharpening. Before modern drills were introduced a wooden drill was used that required great flexibility of hand. This could be used effectively only by women.

Langalanga women selling shell money and jewellery in Honiara.

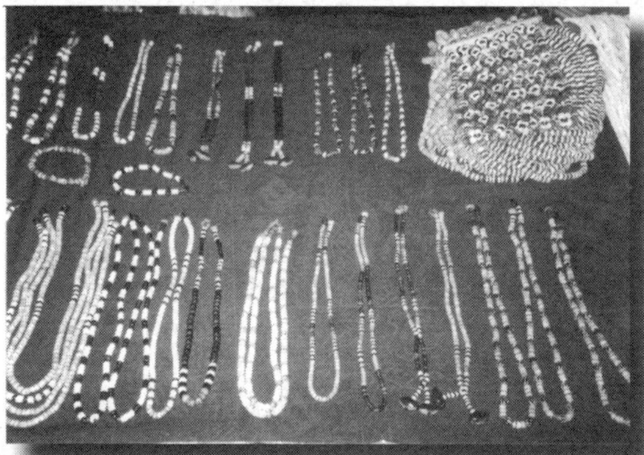

Shell necklaces and shell money on display at Honiara market

The colonization of our islands by the Europeans introduced to our shores new beliefs, values, technologies and ways of living. The traditionally used tools for making shell money also changed. The introduction of the drill and the Chinese stone were notable changes in technology. The modern drill is simple to operate and works more efficiently and effectively, thus raising productivity. It also allows men to perform the drilling task. Today in Kwailao and Lalana most families say that all production tasks are interchangeable among men and women. However, women are believed to be the most highly skilled in cracking or breaking the shells. Women also tend to prefer to do the shaping and heating of shells, whereas men are considered to be exceptionally good at smoothing and sharpening.

In the past, men were always the sellers of the shell money in any exchange. If an *akwala afu* was not sold then the men decided what it could be used for. Today, women are prominent in the sale of both traditional shell money and shell jewellery. Women say that when shell money began to be exchanged for cash, the attitude of men towards the returns of the sale also changed. Men often misused the cash earned from the sales, spending it on liquor or gambling. Today, women have taken over the selling role almost entirely for various reasons but most importantly because they need to secure money to feed their families and to buy more shells from the other provinces.

CHANGES IN SHELL RESOURCE MANAGEMENT

Elders in Kwailao and Lalana can remember when, in their youth, there were chiefs who regulated the harvesting of shells from the Langalanga lagoon according to seasons and occasions. In the past, the chiefs were very knowledgeable about marine territories. They knew the seasons and areas to be protected and for how long the taboos should last. Placing of taboos by chiefs over a reef for a given period of time meant that access to the reef was forbidden for purposes of harvesting of the resources. Taboos could be placed seasonally to ensure a good harvest for the community or to preserve resources for a particular festive event or a culturally important occasion. The village chief would perform a ritual to close the reefs for a period that could range up to several years in length. During this time, nobody was allowed to dive for shells on the reefs. After the period of taboo, the chief would perform another ceremony to open the reefs for harvest.

The act of tabooing a sea area involved killing and burning several pigs as a sacrifice to a god. This was done to ensure a good harvest, to plead to the god to protect the reefs from intrusions, and to ask the god to punish people who violated the taboo. It was said that sharks would fiercely attack any people who violated the taboos. The people were very observant of such restrictions. They had great respect for the chiefs and were fearful of the dangers attached to the taboos.

Today, the traditional rituals involving the sacrificing of pigs are seen as costly and contradictory to the Christian faith, which most village leaders share; therefore the practice is no longer upheld. As one woman noted "People think that to be civilized means to forget culture" (Kwailao household interview #17). Where some memory of a restriction on harvesting persists, it is regularly ignored. "Some people do not observe taboos these days because they are only Christian taboos, not dangerous cultural taboos" (Kwailao household interview # 24). There is generally open access to all reefs.

Some taboos that were imposed in the past were gender specific in that women were not allowed to dive on certain reefs where men could dive. These days, most women, although they may feel reserved, do go diving for shells even on reefs that were formerly forbidden to women.

DEPLETION OF SHELL RESOURCES

New technologies allowed the rate of production of shell money to rise steeply. Over time the demand for shell money for purposes of bride price and compensation payments has risen, as has the price paid for the *akwala afu*. The modern-day acceptance of cash in exchange for shell money makes the trade even more attractive to villagers, who flock into the industry. As the production rate increased, shell resources were depleted, particularly in Langalanga lagoon. Even in the 1970s some types of shell were rare (Connell 1977) and this problem has steadily worsened. Compared to 10 years ago, the shells that one can collect from Langalanga are very small and few and they can only be found far away from the normal collection sites. As a result there is now little diving for shells in the lagoon. The men, who were the more active divers, have lost one of their major roles in the making of shell money. Families have also lost a source of protein, as all the shells used to be eaten, as were other fish and shells that could be collected at the same time. Now that men do not dive they have more time to help the women with production. This in turn increases production and places even more pressure on shell resources.

Now that shells have become very scarce in Langalanga, shells are shipped in from other islands. As coastal ships typically converge on the capital, Honiara, it is to Honiara that women in the villages must travel to buy shell. The people in Kwailao and Lalana purchase shells from Western Province, from Temotu, Yandina and Russell Islands in Central Province, and also from the north and south of Malaita. These are the main suppliers of shells. Other emerging suppliers are Isabel Province and Marau Island in Guadalcanal Province. Shell buyers noted that over the past few years, shells from Yandina have slowly decreased in size and shells obtained from other places are increasingly expensive to buy. They worry that in other provinces the shells will become scarce, as they are in Langalanga.

MARKETING

The people of Lalana and Kwailao travel to Honiara to sell their shell money and jewellery because there are more customers in the capital. Villagers sometimes sell in Auki, the provincial centre of Malaita, and at home, but the selling price is quite low compared to that in Honiara. The shell money and jewellery are taken to the markets strictly for the purpose of cash sales; there is no barter exchange for these items. Most of the buyers are Malaitans, more specifically the people of the northern region of Malaita who have a strong cultural attachment to shell money for bride price and compensation payments. Non-Malaitans who buy large amounts of shell money are usually doing so to pay bride price for marriages with Malaitan women. Other buyers, including foreign tourists, generally purchase only shell jewellery.

SOCIAL UNREST 1999-2002

The social unrest of 1999-2002 was a period of malicious fighting and killing between the people of Malaita and Guadalcanal. During this period there was great suspicion and hatred between the two ethnic groups. The unrest caused enormous disruption to the normal buying and selling of shell money. Movement of people, especially Malaitan women, was restricted for fear of being harassed or even killed. The women therefore could not travel freely to the central market in Honiara, thus it was extremely difficult for them to generate income. When a few Langalanga women did brave the market with their products, troublemakers robbed the market stalls.

Although a lot of compensation was paid during that period of unrest, the bulk of it was in cash, not shell money. People demanded cash compensation because cash is more easily converted into shop goods. Only during the reconciliation ceremonies between warring parties and government was there an exchange of shell money along with cash. Culturally, the exchange of shell money between warring parties is a symbol of peacemaking.

Armed Australian soldier patrolling Auki market, Malaita in November 2003.

After the intervention by Australian police and military, the life of Solomon Islands citizens slowly regained normalcy. There was a boom in the market for shell products as many members of the foreign forces purchased jewellery to wear or to send home as souvenirs. The return of law and order has re-energized the women of Langalanga, who are well pleased with the accomplishments of the Australian-led intervention force.

ECONOMIC AND SOCIAL IMPORTANCE OF SHELL MONEY PRODUCTION

Shell Money Income

Families spend about one month making one *akwala afu* and some jewellery and then the women take the whole lot to market. *Akwala afu* are the main source of income. Depending on the length of the strings, the price for one *akwala afu* ranges from SI$600-$1,000 (US$81-$135)[1]. In addition, women surveyed estimated that they sell on average SI$50 worth of jewellery in a week of marketing.

The women may spend weeks at the market to sell their shell money. Very rarely do people sell their *akwala afu* within the first week of sitting in the market. Usually it takes at least two weeks. When a woman sits at the market for two weeks to sell one shell money and some jewellery for perhaps SI$800-$1,000, this represents her gross income for six weeks. The profits are small because the women must pay daily transport to the market, buy food while residing with relatives in town, purchase raw materials (SI$100-$200 per bag of shells), and cover their boat fare to Honiara and back, labour costs and other miscellaneous expenses. After all these expenses were considered, the women said they usually have a profit ranging from SI$100-$150 for a month.

It would be cheaper to send items for sale to relatives in Honiara but the Honiara-based relatives generally consider it too inconvenient to have to spend weeks in the market to get the items sold.

Use of Income from Shell Money

Being settlers who do not own land, and with marine resources degraded, people of Lalana and Kwailao have little to exploit for income and food. The people depend largely on imported goods and services for their survival, and the shell money business is the main source of cash. The people said that they used their income to pay for what they consider essentials: rice, sugar, tea, tobacco, fresh vegetables, betel nut and clothing. An equally important use of the cash earned from selling shell money is to pay for shells to make more shell money. The expense of travelling to Honiara and purchasing shells is a major drain on people's budgets.

Another small but important use of shell money cash is to pay for labour. There are two reasons why people pay for labour. Casual labour may be required in the process of making the shell money. For example, a group of men or women will be paid to do a particular task. This practice increases productivity and allows for a large order to be quickly filled. Also, when people are so occupied with making shell money that they have no time to make new gardens, they pay for community groups to do their garden work. These are not large farms but small subsistence gardens. Therefore, the investment in gardening labour provides a return in the form of food, but no income.

Houses perched on man-made islets in Langalanga Lagoon.

Rising Economic Needs

Interview respondents, most of whom were in their thirties, noted that there is a greater need for money now than 10 years ago, even at the community level. As the barter system has been largely done away with, everything now must be purchased with money. Basic food items are becoming more expensive and villagers usually have large families as well as extended family obligations. There is an increasing demand from churches, schools and local organizations for contributions, both in kind and in cash. Finally, the purchase of shells is a priority need and also the most expensive one. Unlike in the past, people cannot afford to keep shell money themselves as a form of savings.

As none of these different needs can be ignored, people try to reduce costs by having only small amounts of anything. For example, they buy food but then ration how much they take for a day. Usually, people see it as enough to eat one heavy meal a day, with a lot of tea in between.

Social Impact of the Shell Money Trade on Women

The women of Langalanga face new problems as a result of all the changes in production, demand and use of shell products. The change that has placed the greatest strain on the women is the need to leave their families to travel to Honiara and sit at the city market to sell their products. The women stated that these absences cause their families tremendous inconvenience and stress. Family quarrels may erupt when women gossip about each other to husbands, after they return home from Honiara. Some disputes have resulted in divorce because some women take extreme measures to get their products sold quickly. For example, they will risk their reputations by seeing men in offices to negotiate the sale of their products for special prices.

SURVIVAL OF THE SHELL MONEY INDUSTRY

The people of Langalanga strongly believe that the industry will survive to the moment they pull the last shell from the ocean. Both because people are culturally attached to the craft and because they have few economic alternatives it is not easy for them to stop making shell money, even though expenses rise and returns are meagre. The people do not see any alternative for generating income that is of interest to them.

Although they insist that the trade must go on, the producers recognise their vulnerability and worry about the future of shell resources. They know that the depletion of shells in Langalanga could be followed by depletions in other provinces, meaning that not only must resource management be set in place in their own communities, but somehow there must also be controls across the country.

Some suggest that the industry will only survive if Langalanga people recognize the need to preserve cultural practices and control the commercial use of shell money to safeguard their rich cultural heritage. The average bride price, they note, is increasing and may need to be brought under gov-

SHELL MONEY ESTATE

Mary Geli was a typical Langalanga girl who grew up in the village and had no formal education. She spent her days beside her grandmother observing how she cracked, shaped, drilled and smoothed the shells for shell money making. As she grew up, she made shell money with her mother and grandmother supervising her and giving her practical corrections.

Mary grew up and her family arranged for her marriage. She moved to Honiara with her husband who works in an office. Because Mary was uneducated she could not work in an office so she decided to produce shell money to earn income for her family. Mary started arranging for family members at home to send shells for her to work with. She produced a substantial amount of shell money, which she sold at the Honiara markets. Mary was excited to be able to earn money like her husband, even without sitting in an office.

Mary learned from other women friends that they were planning to go to Bougainville to sell their shell money. She wished to go and her wish came true when her husband gave her his blessing to go. Mary worked hard, both night and day, to prepare 10 shell money strings to take with her. She made two trips and with the money she earned she was able to build 15 three-bedroom houses at Tanagai, on the outskirts of Honiara.

This little estate built from her earnings from shell money was burned to ashes during the ethnic unrest in 2001. To this day, Mary waits with little hope for someone to pay for her loss of property. Now, Mary is a grandmother and rarely travels to buy shells. She spends more time with her grandchildren, and less time on shell jewellery to sell at the market.

ernment control. A rapidly rising population is also of concern, and some women call for access to information on family planning.

Other villagers maintain that the industry will survive if only the harvesting of shells is controlled. Some suggest that Langalanga reefs could be restocked using shells from other provinces but even in this case, there would also have to be some harvest restrictions set up. Also, people note, there need to be regulations and enforcement to stop people from damaging the reef structure, for example with dynamite fishing, and rules to restrict gathering of small shells. Many villagers pointed out that in addition to the traditional chiefs, who have the power to restrict access to the sea, the church also has a role to play. These villagers felt that church authorities should allow traditions to be re-established when they serve to manage resources, and they should also allow rituals to be performed without criticism.

By using taboos or other methods, shells in Langalanga Lagoon will have to be allowed to grow to maturity so that the lagoon will be restocked. Men tend to support the idea of chiefly taboos whereas women are more likely to suggest that the government should enact laws to govern harvesting. The women point out that there are challenges in attempting to reinstate traditional practices because when the old chiefs died, their knowledge of how to conduct the rituals died with them. Also, among the young, traditional culture is being abandoned in favour of modernity.

One of the factors that may handicap the Kwailao community in using taboos is its heterogeneous makeup and the absence of a tribal chief. The people who settled there migrated in from various places. The absence of substantial governance at the community level is reflected in the absence of management of their resources. The resources in and around the area are treated as common property. No one in particular has the right to control access. In Lalana however, the community is homogeneous and governed by a chief who could impose regulations such as seasonal bans.

Clearly there is widespread concern in Lalana and Kwailao, and a great deal of good sense is evident in peoples' suggested remedies. Yet no one suggested that shell money producers on their own could collectively manage the shell resource or adjust their prices or marketing strategies to reduce the pressure on shell stocks. Everyone looked outside of his or her family unit for leadership – to the government, or to their chiefs.

MANAGEMENT OPTIONS FOR THE FUTURE

The study conducted in Lalana and Kwailao clearly identified the severity of the problem of land scarcity for the people. The extreme dependence on the sea coupled with a lack of controls over fishing practices resulted in depletion of their marine resources. The scarcity of local shells means that

THE QUEEN OF SHELL MONEY

Living on an island in the Langalanga Lagoon is a 61-year-old widow by the name of Hilida Geilulu. Hilida makes her living and looks after her family obligations by making shell money.

Hilida's mother and grandmother were both shell money producers. She learned how by watching them. In the past, small girls did not go to school and so they spent more time on shell money production. Hilida was one of them.

Since 1963, Hilida has travelled extensively through Solomon Islands, collecting shells. In 1973, Hilida was living in Small Malaita where, she says, she had a shell money project involving hundreds of women. In 1987, when Hilida was collecting shells in Isabel and Choiseul, the police came to tell her to stop, or else pay for access. She told them she had already paid the paramount chief $100. The police kept demanding that she stop until she said to them, "Look at this $10 note, it has a picture of a Langalanga woman making shell money. Shell money is the basis of the Solomon Islands economy. I am a Langalanga woman and you cannot stop me from harvesting shells". In 1997, an article was published about her in the Solomon Star, and ever since that she is known locally as "the queen of shell money".

the people have lost one of their key sources of protein. The bulk of the shells used in the trade now are sourced from other parts of the Solomon Islands, which has in effect broadened the geographic scope of ecological impacts of the industry.

It is very stressful for the people as they try to preserve their cultural industry amid rising costs of production and resource depletion. They face immediate danger from poverty that involves malnutrition, poor water, disease and over-crowding. With the weakening of the cultural hierarchy of chiefs in the communities, people have abandoned their responsibilities as guardians over their resources. Women also face particular social and financial challenges while trying to market their crafts. These challenges are related to the shift from the cultural use of the shell money to commercial sales of shell money and other shell products. However, sales of shell crafts are now essential to survival in an increasingly cash-based economy. Because of this and also because the people have a strong cultural attachment to the craft and few viable economic alternatives, management of the shell resources is essential.

One option is to reinvigorate the chiefly systems that traditionally governed the communities. Local traditional management approaches can complement regulatory efforts made at the provincial and national levels. As traditional community governance has been weakened by contemporary ideologies, it may well require intervention from some third party to facilitate the expression of peoples' concern and motivate local leaders to act in response. Collaboration among government departments and non-governmental organizations such as the Environmental Concerns Action Network of Solomon Islands (ECANSI) and the Solomon Islands Development Trust (SIDT), is needed to address the issues and problems highlighted in this case study. It will almost certainly require involvement of government departments as well as national, non-governmental organizations to ensure that shells, not only in Langalanga but also in other provinces, are properly managed.

Because women are so heavily involved at all stages in shell money harvesting, processing and marketing, it is clear that management approaches will have to be developed in consultation with women as well as men. Children who labour in the industry also need to be considered. The road to recovery is bound to be very challenging because of the continuing difficulties arising from recent ethnic conflicts. In particular the in-migration of displaced people from Guadalcanal to Malaita places a huge strain on all natural, financial and human resources.

NOTE

1. At the time of writing (2003) SI$1 was equivalent to US$0.135

REFERENCES

Connell, J. 1977. The Bougainville connection: changes in the economic context of shell money production in Malaita. *Oceania* 48:81-100.

Cooper, M. 1971. Economic context of shell money production in Malaita. *Oceania* 41:266-276.

Lewis, A. B. 1929. Melanesian shell money in Field museum collections. *Field Museum of Natural History Anthropological Series* 19(1).60p.

Woodford, C. M. 1908. Notes on the manufacture of the Malaita shell bead money of the Solomon group. *Man* 8(43):81-83.

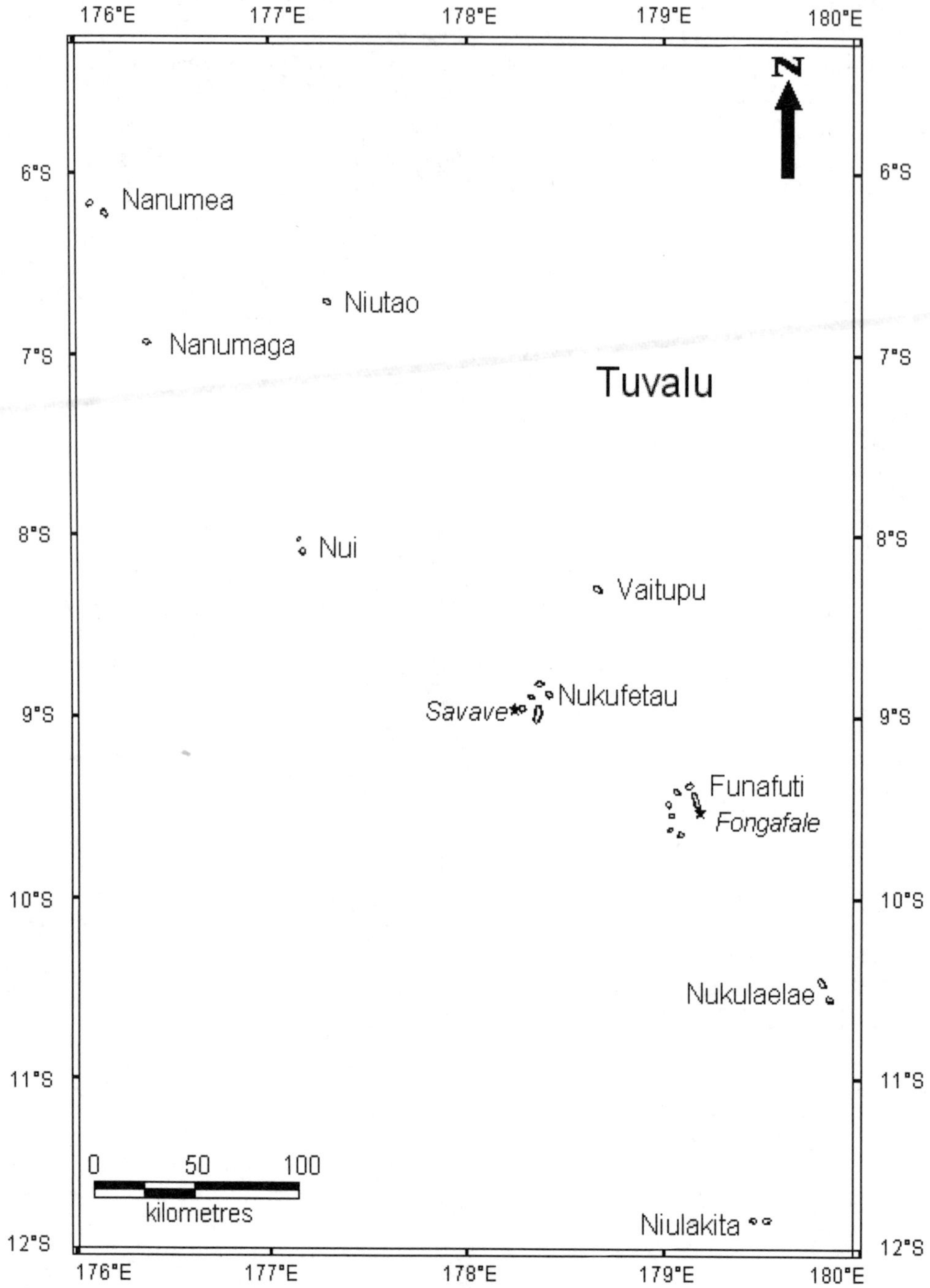

Map 1. Tuvalu, showing location of Funafuti and Nukufetau

4 Seashells On The Seashore: Women's Participation in the Shell Trade on Funafuti and Nukufetau, Tuvalu

Alan Resture & Setapu Resture

"Life is so hard! Fortunately, my income from shell necklaces is supplementing my spouse's income."

— *Siuila Enosa, shell collector, Funafuti.*

"The government needs to look for alternative ways of providing employment and training opportunities for poor people like us women."

— *Tinoi Siliva, Nukufetauan shell collector on Funafuti).*

"I needed the shells for my livelihood. I understand that you have placed a ban on the harvest of pule but to stop me from harvesting shells is like starving me to death."

— *Lina Ielemia, Nukufetau shell collector*

WOMEN'S USE OF MARINE RESOURCES IN TUVALU

Tuvalu is a small, fragmented country and one of the world's smallest and lowest island nations, with a combined land area of only 26 km^2 and elevations of generally less than three metres above sea level. It covers an ocean area of more than 900,000 km^2. The small land area is divided among nine coralline island groups located between 5° and 10° south and 176° and 179° east. This places Tuvalu approximately 1,100 km north of Suva, Fiji, and 250 km south of Kiribati (Resture 2000).

Marine and coastal resources are major drivers for economic advancement in island nations with poor land resources; Tuvalu is no exception. Marine products provide protein and livelihood as well as employment and foreign exchange.

Today, there are a growing number of women who are directly using marine and coastal resources, yet there is little published information on their activities. This case study investigates the role of women in a particular form of fisheries, the ornamental shell trade, which is an increasingly important economic sector. The study looks at the people involved in the

Salateine Vaiafua selling her shell handicrafts in Funafuti.

shell industry, their activities, incomes and economic dependence on the trade. The types of handicrafts made and the marketing opportunities are documented. In addition, we consider which shells are harvested, the state of the resource, and management practices pertaining to the resource.

CASE STUDY METHODS

The case study is based on two islands, Tuvalu's capital, Funafuti, and Nukufetau, the authors' home island. This choice of sites allows for a comparison of activities in an urban setting and a rural one. This is also where the bulk of the shells are collected for Tuvaluan handicraft production.

The study was carried out in two phases. Data collection for Funafuti was in two phases, from 7-14 September and 8-15 December 2003. Interviews on Nukufetau were conducted from 16 December 2003 to 23 January 2004. The key informants interviewed included 24 shell collectors, six private retailers, handicraft centre staff on both atolls, four government officials (Department of Fisheries, Department of Women and island council representatives) and representatives of the National Council of Women (NCW). Four Tuvaluan elders living in Suva, Fiji, were also interviewed between September and November 2003. Personal observations of shell gathering, processing and marketing and informal discussions with target groups were conducted at both study sites. Three questionnaires were used: one for collectors, one for retailers and one for government and NCW respondents.

THE STUDY SITES: NUKUFETAU AND FUNAFUTI

The atoll of Funafuti, with a population of close to 4,500 people, is the only urban area and the capital of Tuvalu. Funafuti has a cash economy that relies heavily on imported goods, most of which come from Fiji, Australia and New Zealand. The main employer is the Government of Tuvalu; the private sector employs only 10% of workers. Artisanal fishing, small food outlets and the sale of handicrafts are the main commercial activities on Funafuti.

Nukufetau Atoll lies roughly 110 km northwest of Funafuti. The population of 630 people is divided among 130 households (Rev F. Ituaso of Funafuti, pers comm 2003), down from a population of 751 in 1991 (Tesfagiorhis 1991). Most of the young men on Nukufetau have migrated to Funafuti in search of employment. At the time of writing, there are only 65 men between the ages of 18 and 49 left on Nukufetau.

Nukufetau's economy is predominantly subsistence, with fishing the main commercial activity. Local fishers sell their fish to the Community Fishing Centre, which sends the fish to Funafuti. There is a women's handicraft centre on Nukufetau which sells shell handicrafts to the National Handicraft Centre in Funafuti.

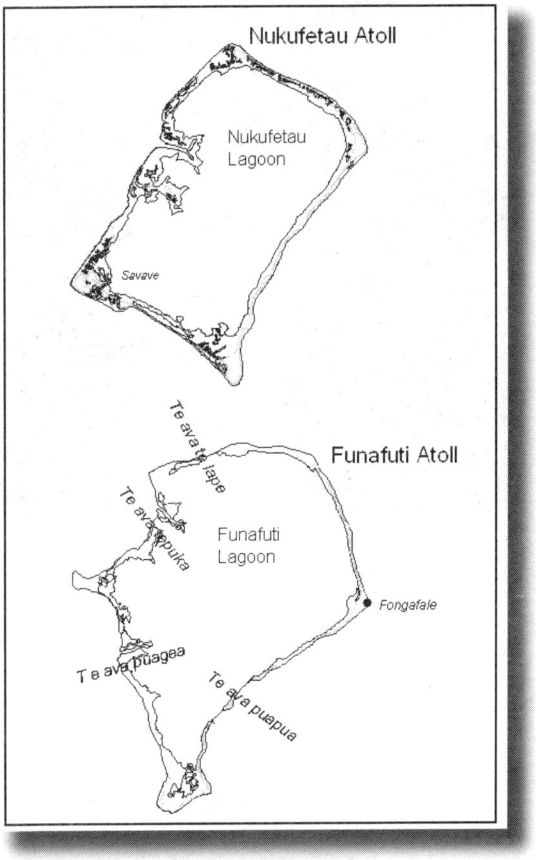

Map 2. Funafuti and Nukufetau Atolls.

THE CULTURAL SIGNIFICANCE OF SHELLS

Shells play a significant role in the culture of Tuvalu. Traditionally, shells have been in use for various cultural purposes. On Nukufetau, valuable family heirlooms (*pakasoa*) have four or five pieces of highly polished mother-of-pearl shell flanking a whale's tooth *(nifo tafola)*, all strung on a bark cord *(lafau)* (Koch 1983). These *pakasoa* are still evident today during the inauguration of chiefs in outer islands and of the Governor-General of Tuvalu. They signify the wealth or status of a family in the community (Feue Tipu, Tuvaluan elder living in Suva).

Usually, fishing masters (*tautai)* acquire pearl shell shanks for making bonito hooks *(pa)* (Passfield 1996). On Nukufetau, once a *tautai* catches

Mr Faimalaga Luka, Governor-General of Tuvalu, wearing a pakasoa.

his hundredth tuna, the *pa* that he used for his catch is given to his *tafaa* (fourth paddler) (Timoteo Seete, Nukufetau fisherman).

The conch shell is also of vital importance to the culture of Tuvalu. It signifies unity. On Nukufetau, the sound of the conch shell is a call for a community meeting or gathering; chiefs and elders gather at the *Falekaupule* (community hall) when they hear the sound. During the festive season (Christmas and New Year), the sound of the conch brings the people together for feasts and traditional dancing.

Young women also enjoy the cultural significance of shells in the form of ornaments during festivities. These are necklaces and pendants worn when they are performing the traditional dance *fatele*. Young women usually go out to collect shells (*pule* and *misa)* on the day of the *fatele* (Talise Tauetia, Tuvaluan elder in Suva). These are threaded together using nylon fishing line to make a necklace.

Women on Tuvalu make a range of handicrafts, including fans, mats, baskets, brooms, vases and ornaments, from different trees and seashells. The simplest handicrafts to make are shell ornaments, of which there are about 25 different types (for example necklaces, bracelets, vases, tablecloths and hair clips). Most shell items are made entirely from shells but others use shells only for decoration.

Today, the culture has extended to giving shell necklaces to relatives and friends leaving Tuvalu, especially when they leave by plane. It is normal to see Tuvaluans leaving the country with more than ten necklaces. These gifts are used to decorate their homes in countries abroad. The use of shells for home decoration is a long-standing tradition. For instance, door curtains made entirely of shells were seen in staff houses of Elisefou School, Vaitupu Island, in the early 1950s (Talise Tauetia, Tuvaluan elder residing in Suva).

Foreign diplomats and government officials visiting Tuvalu also receive shell and other handicrafts during special functions such as conferences and workshops. These are given as gifts to show visitors and guests the Tuvaluan hospitality.

Tuvaluan handicrafts, including a basket trimmed with shells (left). Shell crafts on display (right).

ORNAMENTAL SHELL HARVESTING, PRODUCTION AND MANAGEMENT

Shell Harvesting

There are six main species of shells used in handicraft production, including two species of white cowrie, one black cowrie, two species of land snails, and mother of pearl shell (Table 1). The land snails *(misa)* are the commonest shells used today, especially on Funafuti (Passfield 1996). Mother of pearl is used only for ceremonial purposes such as the inauguration of a high chief or the Governor-General of Tuvalu.

Table 1. Shells used for making ornamental items in Tuvalu.

Type of shell (Tuvalu name)	Latin name	Common name
White cowrie (2 species) (*pule kena*)	*Cypraea obvelata*	gold ring cowrie
	C.moneta	money cowrie
Black cowrie (*pule uli*)	*C. caputserpentis*	snakeshead cowrie
Land gastropod (2 species) (*misa*)	*Melambus luteus*	land snail
	M. fasciatus	land snail
Mother-of-pearl shell (*paa*)	*Pinctada margaritifera*	pearl oyster

Harvesting shells is predominantly the work of women, who generally go out to collect shells two or three times a week. Each harvest lasts an hour or two for both snails (*misa*) and cowries (*pule*). Men and children may also help with the harvest.

Most of the shell handicrafts use *misa.* These shells are found in shady moist areas under coral rubble and debris, usually on the beach. *Misa* is best collected when it is raining, or on nights with a new or quarter moon when shells make their way to the surface of their shelter, making them easier to collect. Collection at other times requires sorting and turning over of coral rubble (Passfield 1996). We observed six collectors on Nukufetau and counted that, on average, women spend at least an hour a day collecting shells, three times a week. Counts of harvested shells showed that women collected about 1,000 a day.

Misa shells are usually collected into a two-litre plastic container or bottle with a lid. After collection, some women kill the *misa* by pouring hot water over them. Others use bleach to remove the odour. Once the *misa* are dead, they are left in the sun for a few days before being processed. *Misa* are abundant on Funafuti, where they are the main shells harvested.

Necklaces (tui misa) made with misa shells.

Pule are collected on the reef flats during low tides. These shells are hardly collected on Funafuti where they are very scarce. On Nukufetau, by contrast, each woman can collect up to 1,000 *pule* per trip. After collection, *pule* are buried for a week to get rid of the meats before they are processed. It is very important that these shells are not exposed to sunlight or hot water as they lose their texture when exposed to heat. Handicrafts made from *pule* are the same as those made from *misa.*

Shell Processing

Sulami Limasene sells crafts from the back of her van.

Once the shells are clean they are transformed into various handicrafts. Making handicrafts from shells requires skill. All shells (*pule* and *misa*) are drilled with a small nail or sharp pointed object. Depending on the type of handicraft, shells may need one or two holes. The different types of shell handicrafts, the materials used in their manufacture and their selling prices are listed in Table 2, with comparative prices from an earlier study by Passfield (1996).

Table 2. Shell handicrafts of Tuvalu in 2004. Note: 1996 prices are from Passfield 1996.

Product	Materials and manufacturing process	Patterns and Price (AUS $) in 1996 vs 2003
Shell necklace • *Pule pepe* (24-inch strand with a butterfly shape made from 80 *pule kena*) • *Tui fafetu* (3 types; 24-inch strands with a star made from 65 small *pule kena* & 5 *pule uli*) • *Tui masela* (pattern from the Marshall Islands) • *Tui pule* (14 types, with triangle or star shapes & one to four strands) • *Tui misa* (14 types, with one or two strands of spiralling shells)	**Basic Materials** • *pule* (cowries) or *misa* (land snails) with either red seeds *(fuaga uli)* or black *misa.* • Nylon fishing line for stringing the shells and seeds. **Optional materials** • Dyed *lau kie* (pandanus leaf) • *taa* (string made from pre-emergent coconut leaves) or polypropylene raffia of different colours. • cardboard cut into various shapes (circles, diamonds, hearts) for *Tui masela* • wool. **Process:** Thread shells and seeds on nylon fishing line. The pule are threaded back to back with a red seed in between pairs.	**1996:** 10 different patterns; price range $0.80 to $2.60. *Tui fafetu* most expensive. **2003:** 33 different patterns; price range $2.39 to $4.49. *Tui fafetu* most expensive.
Pendant • *Malele kena pulatasi* (single flower) • *Malele kena puletolu* (three flowers) • *Malele uli pulatasi* (one flower with black cowrie) • *Malele paa* (mother of pearl bonito hook) • *Malele tapa* • *Malele fou*	**Basic Materials** • *pule* or *misa* shells • 18 inches of nylon fishing line • *taa* or raffia Optional materials • mother of pearl shell • *kolokolo* (coconut string) **Process:** Raffia is wound around the fishing nylon to a certain length and followed by the *misa* or *pule.* The shells are filled with paper so that the line will not slip out. Six shells are placed to each side of the pendant. If *pule* is used to make a pendant the necklace string also has to be threaded with *pule.* The process of making the pendant is the same as for making a brooch (below). To make the *malele paa,* a bonito hook made of mother of pearl replaces the black cowries and the pendant is strung on *kolokolo* (coconut string).	**1996:** 4 different patterns; price range $2 to $3. *Malele paa* most expensive. **2003:** 6 different patterns; price range $3.44 to $4.93. *Malele tapa* most expensive.

Product	Materials and manufacturing process	Patterns and Price (AUS $) in 1996 vs 2003
Hair clasp • *Pine ulu*	**Materials** • *pule* or *misa* shells • cardboard cut into oval shapes • dyed *taa* or raffia or *lau kie* • a small sharp stick made from mangrove / togo *(Rhizophora mangle)* or *gie (Pemphis acidula)* **Process:** The cardboard is wrapped with plain or dyed *taa* or raffia. The shells are sewn to the edge using a needle threaded with *taa* or raffia. In the middle of this oval shape, a bridge is woven from one side to the other. On the woven bridge a small flower like the one used in the brooch is attached. A small stick sharpened with sandpaper is pushed through to complete the hair clasp.	**1996:** $1.80 **2003:** $3.29
Brooch (in the shape of small flowers) • *Pine faka mau gatu* • *Pine gatu pule* • *Pine faka mau gatu tio taa*	**Materials** • *misa* or *pule* • nylon fishing line or white *taa* • small safety pin. **Process:** Strands of white *taa* are wrapped around a small red seed until the seed is covered. The *misa* or *pule* with a hole at the tip, are sewn around the edge using a needle threaded with *taa* or raffia. After that a small safety pin is attached at the bottom of the flower.	**1996:** $0.80 **2003:** $3.14–3.44
Head band • *Fou pule* • *Fou laga misa* • *Fou laga pule*	**Materials** • *Misa* or *pule kena* • cardboard • pandanus leaf • *taa* or *raffia* **Process:** Stiff cardboard is cut into strips and wrapped with fine pandanus leaves. The plain or dyed *taa* or raffia is woven into the band following a certain pattern. Then *misa* or *pule kena* (with a hole at the tip) are sewn with raffia thread to the edges of the headband.	**1996:** $1.70 **2003:** $2.99-5.23

Product	Materials and manufacturing process	Patterns and Price (AUS $) in 1996 vs 2003
Door curtain • *Pui mataloa pule* • *Fakavasa ki misa*	**Materials:** • *Pule* or *misa* • nylon fishing line • a 2x2 timber. **Process:** Shells threaded on strands 190cm long are tied to nails that have been hammered into a two-by-two timber. For a door curtain there are about ten or twelve strands.	**2003:** $46.64
Belt • *Fusi kolokolo pakolo pule lasi, ite kogaloto, filiki* • *Fusi misa* • *Fusi pule*	**Materials** • *misa* or *pule* • Woven pandanus leaves **Process:** Shells are sewn to the edges of the woven belt for decoration.	**2003:** $3.59-12.41
Bag • *Kapa puatolo*	**Materials** • *Pule* or *misa* • nylon fishing line	**2003:** $14.30-17.80
Napkin ring • *Mea fao sologutu pule*	**Materials** • *Pule* or *misa* • nylon fishing line	**2003:** $6.43
Wrist band • *Taulima pule*	**Materials** • *Pule* or *misa* • nylon fishing line	**2003:** $1.94
Vase • *Teu pule* • *Teu lasi magamaga pule* • *Teu foliki fakatu peni pule*	Materials • pule and fishing nylon. **Process:** The strings of *pule* are carefully and skilfully sewn together in the shape of a vase.	**2003:** $6.62-20.48
Sets	**Materials** The sets consist of a pair of ear rings, a brooch, a hair clasp and a *malele* pendant.	**2003:** AUS $14.30-17.80

The State of Shell Resources

Anecdotal evidence suggests that *pule* are scarce around the main settlement of Fogafale on Funafuti. Both *pule kena* and *pule uli* are fast disappearing. A boat is therefore needed to go to uninhabited islets to collect shells, making it hard for women to collect them.

On Nukufetau, both *pule kena* and *pule uli* are abundant, so these are the main shells used for making necklaces. Women go out about 10 metres from the beach to collect *pule kena*, whereas *pule uli* is collected farther out on the edges of the reef. *Misa* are also abundant but hardly harvested because the selling price is relatively low.

Level of Exploitation

One of the six private retailers we interviewed reportedly sold 30-50 necklaces a day, three days per week. Assuming on average 200 shells per necklace we estimate that her daily sales account for roughly 10,400 shells. The weekly sale of 150 necklaces is therefore equivalent to 10 kg of shells. Consequently, 520 kg of shells may leave the country in one year, from any one retailer. At this rate, given that the number of retailers is limited, it seems likely that the status of the resource at a national level is not threatened. However, caution still should be practiced to avoid local over-exploitation of the resource, such as has occurred with *pule* on Funafuti.

Resource Management Practices

The government of Tuvalu has no regulations pertaining to the collection of shells. No harvesting licence is required, mainly because the government has not seen any threat to the resource. Because there are no formal management tools in place for shells, collectors feel free to collect any amount. Collectors on Funafuti said that there were sufficient shells, especially land snails (*misa*), and management of the resource was not needed. It was interesting to note that all the interview respondents believed that the more you collect *misa*, the more they reproduce.

Traditionally, the chiefs and elders determine the allowable harvest for local marine resources. In the case of shells, the *Kaupule* (island council) has traditional authority to determine the quantity to be harvested. On Funafuti, the council has placed a verbal ban on shell collection by anyone who is not native to Funafuti.

On Nukufetau, the harvest of cowrie (*pule*) is seasonal. The *Kaupule* (island council) with the endorsement of the *Falekaupule* (council of elders) determines when to close and reopen the harvesting season. In 1995, these councils placed a verbal ban on the export of unworked cowrie shells from Nukufetau but they did not impede the export of handicrafts made from these shells (Passfield 1996). In spite of the ban, Nukufetau women on Funafuti continue to receive unworked shells (*pule*) from relatives in Nukufetau.

If apprehended, offenders on both islands may be called to the *Falekaupule* (meeting hall) to receive counselling from the elders. This is a form of traditional punishment, as it is embarrassing to be called in by the elders for counselling. However, enforcement is typically not effective; the traditional authorities have no actual legal right to forbid collectors from collecting shells. Fisheries officials feel that there is a need to work closely with the Funafuti *Kaupule* (island council) to develop formal bylaws relating to the management of shells (Nikolasi Apinelu, Fisheries Extension Officer; Tihala, Deputy Director of Fisheries Department).

Alternatives for Management

As noted above, *pule* (cowrie) are still abundant on Nukufetau, but are fast disappearing from Funafuti. The land gastropods are still abundant on both islands. One alternative for management is to have a bylaw restricting the harvesting of shells, for example closing particular areas or imposing a closed season. Establishing the best time for seasonal closures would require an understanding of the breeding season. Scientific study is needed to establish when shells are most vulnerable. To restrict the harvesting of small shells may not be applicable since the trade depends on big, mature shells, meaning that the undersized shells are never harvested. If the resource did approach depletion, an argument might be made to restrict harvesting of the largest shells, if these were shown to be the most fecund breeders.

The other alternative is an economic one: to license the collection of shells. A licence for collectors could allow them to collect up to a certain amount – a quota. Once the quota was reached, collectors would have to stop collecting. If a licence holder could not collect her quota, she could sell that quota to another collector.

Although women use shells every day, they are not actively involved in resource management. Women have great potential to take the leading role in the decision-making process pertaining to the management of shell harvesting. Their experience and knowledge would be a valuable asset to a management institution. The key constraint on women's participation in management is cultural. Owing to cultural taboos, women are not allowed to voice their concerns in the *Falekaupule*.

IMPORTANCE OF THE ORNAMENTAL SHELL TRADE TO WOMEN'S LIVELIHOODS

In Tuvalu, women have limited economic alternatives available to them. Whereas men can often find full-time or seasonal jobs, women must work harder and be more creative to find economic opportunities. They may sell food to workers on Funafuti, sell candies and ice blocks to school children or sell handicrafts. Of 19 people involved in the shell trade who were interviewed at length, only two were men. This reflects women's dominance of the shell trade.

While traditional shell handicrafts have been produced since at least the early 1900s (Talise Tauetia, elder), the commercial shell trade only began after the separation of Tuvalu from Kiribati in 1973 (Lina Ielemia, Nukufetau handicraft producer). One respondent on Funafuti started

Shell crafting couple Sulami Limasene and Limasese Olikene, showing off their wares.

handicraft production 25 years ago with the sale of shell necklaces, mats, fans and hair clips. Another woman began making handicrafts, including shell necklaces, 30 years ago. Most respondents, however, became involved within the past five years. We were told that some of the women who had been involved in the trade on a longer-term basis had migrated to New Zealand.

On Funafuti there were 13 women involved in commercial shell handicraft production at the time of the study (2003). Most of them had migrated to Funafuti from Nukufetau Island. They were selling their handicrafts to six retailers outside the airport. On Nukufetau, there were only six producers active in 2003.

The women interviewed for this study included harvesters (many of whom also manufactured shell crafts) and private retailers. They ranged in age from 24-65 years and all were entirely dependent on the shell trade for personal income. There were also two men who helped their spouses with the collection and production of shell handicrafts. One man depends entirely on the shell trade for personal income; the other supplements his regular income from government employment.

For most of the shell collectors, shell sales provide between 50% and 70% of total household earnings. On average, collectors earned AUS$100 a week while retailers on Funafuti earned AUS$185 a week. This was the net profit after expenses. We counted 24 harvesters active on Funafuti and Nukufetau. Assuming an average profit of $100 each, we calculate that they earn a total of $2,400 per week, or $124,800 per year. The total annual income (after expenses) for the six retailers was calculated to be about $57,200. In addition to this are the sales that went through the National Handicraft Centre (NHC) on Funafuti, estimated to total AUS$29,053 in 2003. The value of NHC sales has increased over the past decade. In 1995, sales of shell handicrafts reportedly earned only AUS$5,130 (Passfield 1996).

As these figures show, the shell trade is economically important, not only to the individual women but also to the national economy. It is also a growing trade, yet the shell trade is not counted as part of the gross domestic product of Tuvalu. The government does, however, recognise the shell trade as commerce. Before they are allowed to sell handicrafts, women must buy a government licence, which costs AUS$100 and is valid for a lifetime. In addition, the island council of Funafuti requires women to pay AUS$100 per year.

LIFE IN THE SHELL TRADE: LINA IELEMINA'S STORY

Lina Ielemina, age 64, was born on Nukufetau Island. She has been making seashell handicrafts since 1969. She learned how to make handicrafts from her cousin when she visited her on Funafuti Island.

Lina Ielemina

Lina was married and had six children. Because her husband was unemployed, the family's only source of income was from their nephew, a seafarer. Life was very hard, especially after her husband died in 1990. Because she needed money to meet the children's school fees and daily needs as well as community and church contributions, Lina started making handicrafts for sale.

The handicrafts Lina made out of seashells were belts and various necklaces, each of which had a special name: *tui pule tuatasi, tui pule tualua, tui pule tuatolu, tui pule tuafaa, tui pule tualima,* and *tui pule tuaono.* Her children and her husband helped her in collecting *pule.* When she first began there was no formal market so she sent her handicrafts to her niece in Nauru, where they were sold for AUS$1 each. The money from the sales was used to buy things they wanted from Nauru. Nowadays the types of handicrafts she once sold to Nauru do not exist in the market.

At one point, traditional authorities placed a verbal ban on collecting shells. Lina was taken to court with some other women who were caught collecting seashells. All the women who were taken to court said that they were guilty, except for Lina. She explained to the court that she needed the shells because they were the source of money for her family. She understood that the elders of the island had placed a ban verbally, but there was no bylaw. The court dismissed her case because it also did not recognise verbal bans.

Today Lina and her two daughters are still actively involved in the shell trade. She collects her shells during the day at low tide or during neap tides. Once the shells are cleaned the women make their necklaces. These are then sold to the local women's handicraft centre. The money Lina receives from her handicrafts is used to support her family. She had a problem with the local handicraft centre because payment was usually one to two months late. Lina now has a trading partner who orders cowrie shell necklaces from her. She hopes to find a market overseas where she and other Nukufetau women can sell their products directly and at a good price.

Incomes that women earn from the shell trade are used for the sustenance of families, women's group contributions and community contributions. Collectors on Nukufetau indicated that at times all that they earn goes to either women's group contributions, church contributions and/or commu-

Selling shellcrafts in Funafuti

Tautala Pusineli strings shells while awaiting a customer.

nity contributions. On Funafuti, about 50% of collectors' incomes from shell handicrafts goes to such obligations. In contrast, incomes earned by male members of Tuvaluan families typically go to other commitments such as repayment of loans. Loans are usually for their bicycles, motors, fishing gear, building materials and major family contributions such as payments for funerals and weddings.

MARKETING CHANNELS FOR SHELL HANDICRAFTS

Current Conditions

There are few marketing opportunities for handicrafts in Tuvalu. Women on Funafuti who collect shells and then process them into necklaces and other handicrafts may sell them to the National Handicraft Centre (NHC), which handles the greatest share of the shell handicrafts of Tuvalu. Alternatively, women may sell to smaller retailers. Products are sold mostly to departing travellers. Because the number of tourists that visit Tuvalu is small, the main buyers are Tuvaluans leaving the country. Special events can induce a sudden spike in sales. For example in 2003, October alone brought in a gross revenue of AUS$6,063 to the NHC (Tumu Salili, salesperson). This was because of the 25th anniversary of independence, when the demand for shell necklaces increased sharply in response to increased visitors from abroad.

Once the NHC considers that it has enough necklaces to satisfy immediate demand, it stops buying from the producers. Women then must sell their necklaces to private retailers who operate small market stalls. The private retailers are also women. Because private retailers pay more than the NHC, Funafuti collectors/producers actually prefer to sell their handicrafts to them. The NHC is therefore more dependent on handicrafts sent in from producers who live on the outer islands.

Women on Nukufetau are reluctant to make handicrafts for the NHC because the market is so small and irregular. Handicrafts on Nukufetau (mainly those made from *pule*) are sent to Funafuti only when the supply there is not able to cater for the demand. Demand fluctuates because the number of travellers determines sales of handicrafts. The peak season is from November to January when students come and go.

An added difficulty for Nukufetau producers is that payments for shells sent to Funafuti may not arrive until a month or two after the handicrafts are sent. Also, the women's handicraft centre on Nukufetau suffers from financial constraints and so is looked after by a volunteer. Because of these difficulties the number of women on Nukufetau making shell necklaces and mats dropped from 11 in 2002 (Economic Research and Planning Division, pers comm 2003) to six in 2004.

Potential Markets

There is potential for increased marketing of shell handicrafts abroad. However, there are constraints that hinder the flow of goods to overseas markets. The market for shell handicrafts is not secured, in part because the National Council of Women on Funafuti assumes that there are not enough shell handicrafts being made to cater for the demand from abroad of 10,000 necklaces per year (M. Paeniu, Tuvalu NCW). Indeed, as seen above, the delay in payments for necklaces has discouraged women on Nukufetau and some have given up making handicrafts even though cowrie shells are abundant.

Banner, fans and other shell crafts on display.

Our research indicates that, in theory, the overseas demand could easily be met. Our respondents estimated that a woman can make 10 necklaces per day in her spare time. Assuming 200 shells per necklace and a harvest rate of 500 shells per hour, a woman would have to work four hours to collect shells for the 10 necklaces – a not unreasonable workload. A work force of 20 harvester-producers, each making 10 necklaces a week, would meet and exceed the demand for 10,000 necklaces per year. Given the incentives of reasonable prices and a steady market, it is likely that they could double this level of production without difficulty.

Tuvaluan shell necklaces come in many varied patterns.

To improve their businesses, women in the shell trade need

training. No collector or retailer has been trained in how to run a small business. They have limited knowledge of bookkeeping and other essential skills.

ENSURING A FUTURE FOR THE SHELL TRADE OF TUVALU

The shell industry appears to have potential for expansion and is a possible engine for economic development of women in Tuvalu. However, some constraints need to be addressed. For example, women are expected to make community and church contributions, which drain family economic resources. If women were free to devote their incomes to their families, they could improve their social welfare. Community leaders should understand the need for women to focus on their family affairs before they can contribute to community activities.

The economic welfare of women is an area that requires government attention. Whereas some women manage to find paid employment, many have very limited opportunities. Women have been given scant attention in the past but it is now being realized that through their reproductive, household and informal sector work, they contribute enormously to Tuvalu's economy. The government therefore, through its Department of Women's Affairs, should consider how to provide employment opportunities for women so that they will have more options to meet their economic needs.

An area of weakness in the marketing of shell handicrafts is the poor documentation of sales. Because there was no database we could not accurately determine the number of necklaces sold by the NHC and other retailers. The marketing centre does not have a computer. All sales are recorded manually and the accuracy of the records, such as they are, is unknown. Knowing how much and to where shells are being exported is important for developing marketing strategies. Knowing the total amount of shells sold is also necessary for the management of the shell resource.

Training is an important requirement of the shell industry and should be on the list of priorities for the National Council of Women. Women who are involved in the shell trade need training so that they are more equipped to enhance the profitability of their small businesses. Training is also needed to equip women to participate in resource management. If they are more aware of the biological cycles of shells, then they can harvest at a more sustainable rate so that the environmental returns (shell growth and reproduction) will in turn sustain their trade.

There is a need also to assess the environmental state of the shell resource. A biological study is recommended to scientifically test the collectors' theory that the more one collects land gastropods the more they reproduce. A study on how to get the *pule* on Funafuti to recover is also highly recommended. Once an assessment is complete, sound management practices can be implemented.

Research on the biology of shells is necessary and requires government support. The Marine Studies Programme of the University of the South Pacific (USP) is an ideal institution to carry this out and optimally this would be in partnership with the Fisheries Division of the Government of

Tuvalu and the *Kaupule* of Funafuti. In this way, capacity could be built in Tuvalu and future studies could be carried out by locals. We recommend that both a biological study and a socio-economic study be conducted concurrently. Management measures rely on the outcome of the biological study, but regulations will only succeed if the social and cultural context is well understood and taken into consideration.

As far as management is concerned, women must be actively involved in the policy-making process because they are custodians of the resource and have a wealth of knowledge on the status of the shells that they use. Harvesters also need to be kept informed of policies that will affect their economic well-being. One way forward would entail having women's representatives included in island councils and, through the Department of Women's Affairs and the National Council of Women, in government decision-making bodies. If these representatives were active and vocal, there would be hope for women's involvement in the decision-making processes.

One of the main constraints on the shell trade that we noted is that the various government and non-governmental stakeholders fail to communicate effectively with one another. While there are various institutions in place, their functions are not clear and there is no integration among them. The National Council of Women does not work closely with the Government's Department of Women (Asita Iakopo, Department of Women; Malama Paeniu, National Council of Women). The Fisheries Department is also not working effectively with the island councils on issues pertaining to the management of the resource. While efforts by both parties are underway, the process is very slow.

To conclude, this study has shown that women can reap significant economic benefits from the shell trade, provided proper arrangements are made for marketing the shell products. Once a good market is secured, more income will be generated, thereby improving women's status in Tuvalu. However, there is also need for caution because, as seen in the case of the shell money trade in Solomon Islands (Fidali-Hickey & Whippy-Morris this volume), local depletion of shell resources and possibly even depletion on a national scale, is possible in the absence of sound management.

REFERENCES

Koch, G. 1983. *The Material Culture of Tuvalu.* Suva: Institute of Pacific Studies, University of the South Pacific.

Passfield, A. 1996. The uses of shells in traditional Tuvaluan handicrafts. *SPC Traditional Marine Resource Management and Knowledge Information Bulletin* # 7 September 1996.

Resture, A. 2000. *Te Falekaupule o Tuvalu: Analyzing Resource Management Practices in Tuvalu.* Unpublished AOC paper, University of Hawaii, Honolulu.

Tesfagiorhis, H. 1991. *Tuvalu: 1991 Population and Housing Census.* Analytical Report, Government of Tuvalu, 2: 1-5.

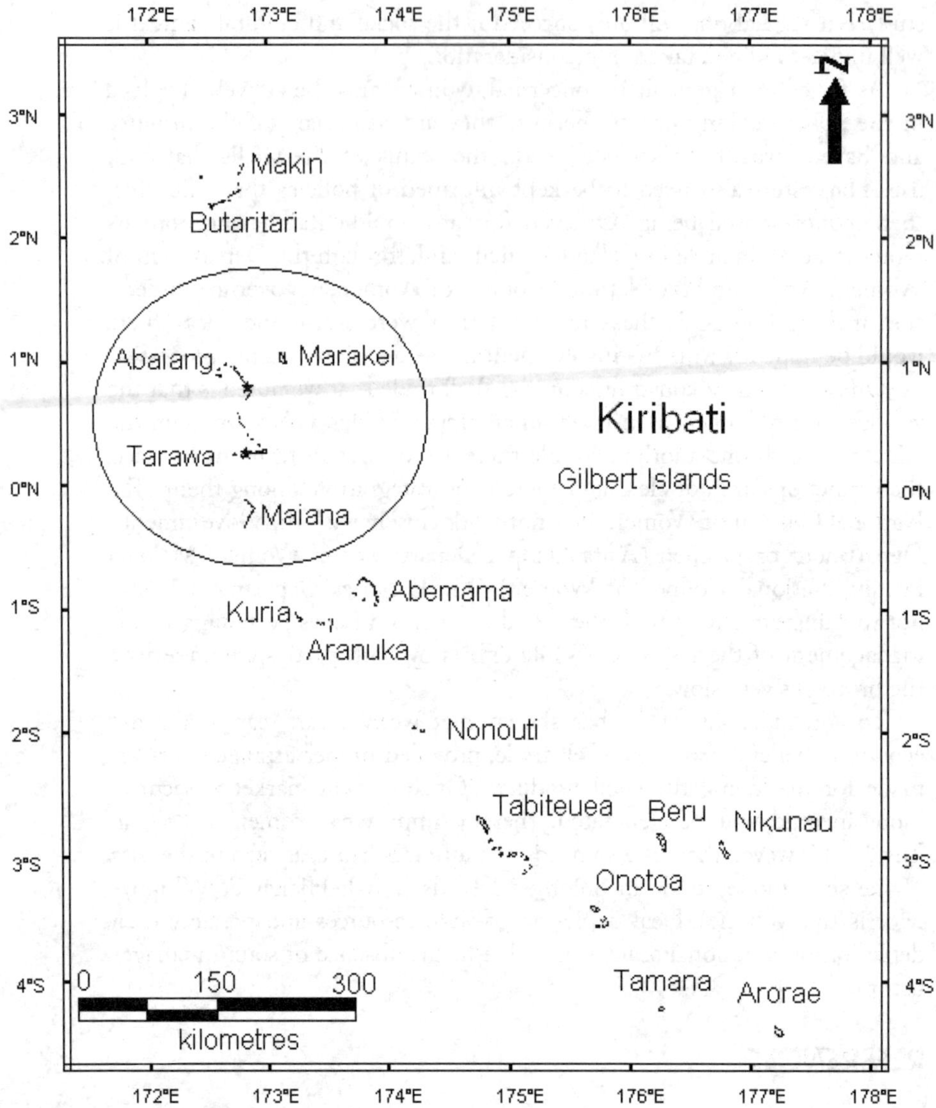

Map 1. Kiribati, showing location of Abaiang and Tarawa atolls.

5 Women And The Anadara Fishery: A Case Study In South Tarawa, Kiribati

Lilian Fay-Sauni & Samasoni Sauni

THE ANADARA FISHERY OF KIRIBATI

In many countries in the South Pacific, women are major contributors to the fisheries for shellfish and other invertebrates. For example, women harvest the clam *Anadara holoserica* from intertidal areas and in shallow waters of Kiribati while men concentrate more on the deeper water shellfish beds (Tebano, in prep). In South Tarawa, the extent of the women's involvement in the harvest of this resource varies according to whether they collect only for household consumption or also for small-scale commercial purposes (Paulay 1995, Tebano 1990).

Lately, there have been reports of an increasing strain on the marine resources in the South Pacific (Vunisea 1997). In the case of South Tarawa, women harvesters have recently noted reductions in density and shell length of clams (Temakei Tebano, USP, pers comm). Some factors that could affect the number and size of *Anadara* include fishing pressure, habitat change, fecundity, recruitment success, and natural mortality. These influences and others, such as pollution and species competition, need to be carefully examined.

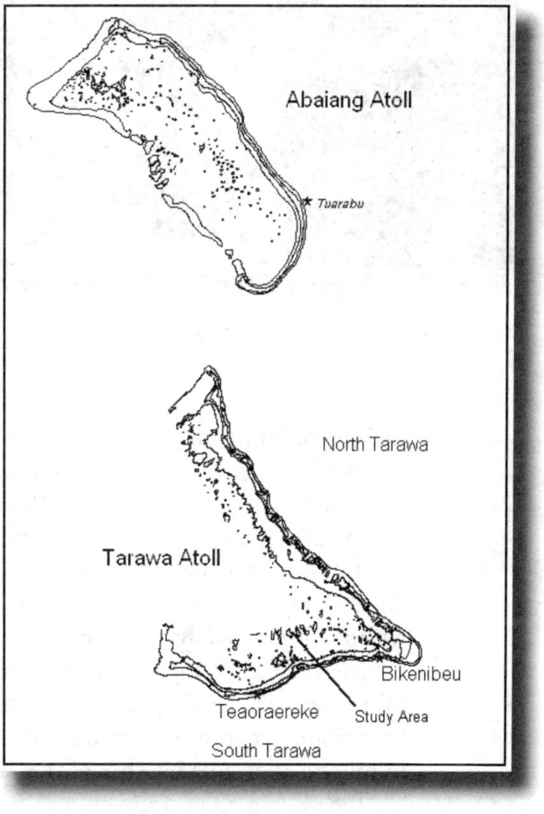

Map 2. Tarawa and Abaiang Atolls in the Gilbert Islands of Kiribati

Anadara holoserica plays an important role in the lives of many women and their families in Kiribati. Some families rely solely on these clams for protein, therefore any decrease in density and size is a cause for concern. Hence, it is important to determine the impact of fishing pressure on the resource so that, if needed, practical means to conserve and manage this resource can be put in place. The goal is to would avoid catastrophic depletion of stocks and associated social problems.

Sieving juvenile Anadara from sediment.

For this case study we measured the density and size structure of *Anadara* in a heavily fished area of South Tarawa and compared these to data from an undisturbed clam population on neighbouring Abaiang Island (Map 2). We use the data to test the hypothesis that there are no significant differences between clam populations of South Tarawa and Abaiang. If the hypothesis cannot be proven, this suggests that real differences do exist, which may be related to women's fishing activities.

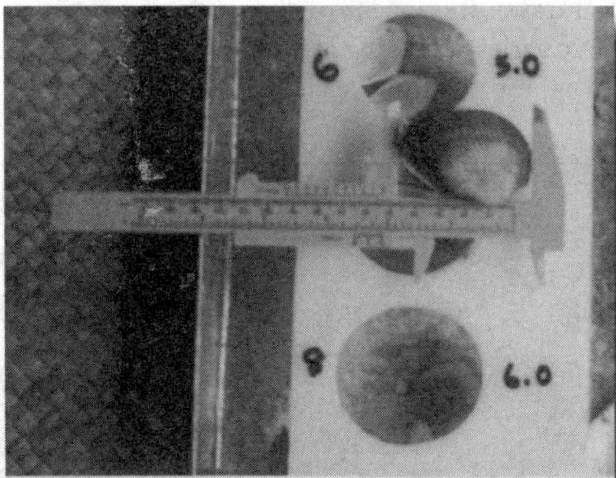

Measuring the length of an Anadara shell.

CASE STUDY METHODS

Surveying the Clam Beds

Fieldwork was conducted on two study sites (Tarawa and Abaiang) from February to June 1999. Sampling within the study sites was done by counting and measuring all Anadara clams inside randomly placed quadrats (0.5m x 0.5m in size). This method provides an estimate of the clam size and density over the whole study area.

To assess how the population changed with distance out from the beach, two quadrats were placed haphazardly at 15 metre intervals along straight line transects that stretched from the beach to the lagoon. All sampling was done during low tide. One set of samples (one transect) could be completed within approximately two hours.

The time and budget available were limited. We therefore did a statistical analysis of the first couple of transects to determine how many would be required to properly characterise each site. We then carried out the minimum allowable number of transects, which was three per site.

The placement of transects was random and was determined using maps prior to arriving on site. Transect lines were laid out from the lower shoreline across the sandflat toward the lagoon (ca. 300-700m). Tebano (1990) has used this technique and this quadrat size on seagrass and invertebrates.

Anadara visible inside each quadrat were counted before the substratum was excavated down to 20 cm (or to bedrock level). The sediment was sieved through a 2 mm mesh to catch any juveniles. All *Anadara* encountered were placed in labelled plastic bags and taken to the laboratory where their size (shell length) was measured.

Analysis of the Data

Analyses were performed to determine the population structure of *Anadara* in each study site, and to make comparisons between the two sites. All data were entered into a Microsoft Excel database, standardized and tested for normality and homogeneity before being subjected to further analysis. The average density (number per square metre) and the average length of *Anadara* were calculated for each transect. Data on shell length were categorized into five size classes (age groups) before analysis using Chi-Square tests. Analysis of Variance (ANOVA) was also used, to test for differences within and between the two study sites. For this we employed the Statistical Package for Social Sciences (SPSS) software.

KIRIBATI FACTS AND FIGURES

The Republic of Kiribati is located in the Central Pacific Ocean between 4° N and 3° S latitude and 172° and 177° E longitude. Kiribati has three island groups widely dispersed over 3 million km^2 of ocean: the Gilbert Islands, Line Islands and Phoenix Islands. The atolls of Kiribati, which seldom rise more than 5m above sea level, are mostly composed of coral debris and frequently consist of a thin, broken ribbon of land partly enclosing a lagoon.

Soil on the atolls is generally less than a metre thick and overlies several metres of sand. It supports little growth apart from seaside scrub, pandanus and coconuts. The lifestyle of the people has consequently been very focused on the sea. A self-sufficient, subsistence economy revolves around fishing, coconut trees and the harvesting of staple root crops (Douglas and Douglas 1994). The country has limited potential for further development of its agricultural base or manufacturing industries but could increase yields from marine resources.

In 1987 the United Nations designated Kiribati as a least developed country. GDP per head was AUS $696 in 1989 (Douglas and Douglas 1994), which is one of the lowest in the Pacific Islands.

TARAWA ATOLL

Tarawa Atoll is an L-shaped chain of islets enclosing a wide, shallow lagoon that opens to the west. The atoll lies in the Gilbert group (Map 1) and is divided into North and South Tarawa, with the administrative and commercial center of Kiribati on South Tarawa (Map 2). The more

Tarawa lagoon and sand flats.

rural North Tarawa, considered a distinct municipality, corresponds to most of the eastern rim of the atoll. It is separated from urban South Tarawa by shallow passages, some of which have been closed by causeways. South Tarawa includes the islets along the southern rim of the atoll. These islets are inter-connected by causeways, allowing for easy commuting and employment opportunities.

According to the 2000 census, Kiribati has a total population of 84,494, 43% of whom reside on South Tarawa (Kiribati Government 2002). The population is especially concentrated in the busy and populous islet of Betio. The large population of South Tarawa is primarily the result of urban drift from neighboring islands; native Tarawans are in the minority. Since World War II, the population has increased by 1,700% (Kiribati Government 1996). From Teaoraereke to Buota on South Tarawa, where the research site was located, women make up 49% of an overall population of 15,761 people (Kiribati Government 1996).

On the eastern (ocean) side, Tarawa has a largely intertidal reef platform several hundred metres wide. On the western (lagoon) side there is a broad sand flat. Most of the western rim of the atoll is below sea level. This submerged barrier reef lies at a depth of a few metres and is interrupted by a single deeper passage near its southern end. The lagoon of Tarawa, which covers an area of 350 km^2, supports diverse and abundant shellfish stocks (Chen *et al* 1995; Kimmerer and Walsh 1981).

Tarawa lagoon is shallow (mean depth 6 m, maximum depth 25 m) and the mean tidal range varies between 1.5 and 2.2 m. The shore consists of a sloping sandy beach 20-30 m wide, limestone rock outcrops and mangroves. Beyond the beach are intertidal to shallow subtidal sand flats that are 0.3-2.5 km wide. These flats can be subdivided into the shore (exposed at all low tides), the mid-intertidal (exposed only during the lowest tides of the month, called spring tides) and the lagoonward area (exposed for a short time only during the lowest spring tides). Along the lagoonward edge of the sand flats there are extensive seagrass beds. Poorly sorted and coarse sands dominate the substrate, and there is abundant gravel derived from broken shells and coral fragments (Paulay 1995).

The South Tarawa Study Site

The study site on South Tarawa stretched along the intertidal flat from Teaoraereke village to Bikenibeu (Map 2). *Anadara* is most common in the seagrass beds adjacent to the village of Ambo and extending to Bikenibeu Village (Tebano and Paulay 1995). This study area was selected because initial surveys revealed that women were fishing there frequently and *Anadara* were relatively abundant.

The sand flats were composed of coarse sand and coral fragments covered with filamentous green algae. Fine sand and silt occurred mainly in the turtle grass beds (*Thalassia hemphrichii)* along the outer edge. *Anadara* was present in the turtle grass beds where the water is always at least 30 cm deep.

ABAIANG ATOLL

The Abaiang lagoon (Map 2) has physical and biological characteristics that are very similar to those of Tarawa lagoon. The atoll is located just north of Tarawa, and is separated from Tarawa by a channel a few miles wide. *Anadara* is generally found among turtle grass that grows along the outer edge of the intertidal flats and extends into the lagoon. The intertidal sand flats are approximately 500 m wide, while the average width of the turtle grass beds is about 150 m. The sediment in the sand flats ranges from coarse sand along the shore to fine and soft silt along the inner edge of the lagoon.

On Abaiang *Anadara* is collected mainly by women, but in some instances school students also collect it.

The Abaiang Study Site

The study site was located along the intertidal flats adjacent to Tuarabu Village (Map 2). The distance from the shore to the lagoonward margin of the intertidal flats is approximately 400 m. Small patches of turtle grass and seaweed (*Halimeda* species) are found throughout the intertidal zone, with increasing density of *Anadara* towards the lagoon. The substratum is mainly fine sand and coral rubble near the shore. Fine sand and silt are more common towards the lagoon. *Anadara* were found either half-buried in the sediment or aggregated in greater numbers in the denser patches of turtle grass.

This site was chosen because of the abundance of *Anadara* as determined from preliminary surveys. Also, local residents provided the information that *Anadara* gathered from this particular area tasted bitter and consequently, women did not collect them. Hence, the site represented an area of low *Anadara* fishing pressure.

It was assumed that because the two lagoons are very similar morphologically but vary greatly in terms of fishing pressure, a comparison of the *Anadara* populations in the Abaiang and South Tarawa sites would reveal the impacts of fishing on this resource.

ANADARA POPULATIONS ON TARAWA AND ABAIANG

Comparison of Clam Density

The average number of *Anadara* per square metre on South Tarawa transects (2.9 clams m^{-2}) was lower than that of Abaiang transects (5.2 clams m^{-2}). Analysis of Variance confirmed that the difference between the two sites was statistically significant ($p=0.03$). On South Tarawa, the average number of clams was the same for all three transects. Average clam density in Abaiang was more varied but the variation among the Abaiang transects was not statistically significant.

Comparisons of Shell Length

The average length of *Anadara* at the Abaiang site (50.7 mm) was greater than that of urban South Tarawa (38.2 mm).

The two sites also differed in terms of the size distribution of the *Anadara* populations. In the less harvested site (Abaiang), 41-50 mm long shells were much more abundant than other size classes but very small shells (0-30 mm) were not found (Graph 1). In the harvested site (South Tarawa) the size distribution was much more regular. Juveniles were present but no size class was particularly dominant.

Examination of more fine-grained size distribution patterns (Graph 2) shows that the pattern in Abaiang exhibits a number of distinct peaks and troughs. The pattern in Tarawa, by contrast, is relatively flat and it is difficult to distinguish definite peaks. Peaks in this kind of graph are interpreted as evidence of sporadic recruitment into the population. A large number of larvae settle at one time and as they grow they can be identified as an abundant cohort of a particular size. The space between peaks represents a space of time between episodes of spawning and larval development. If in comparing the

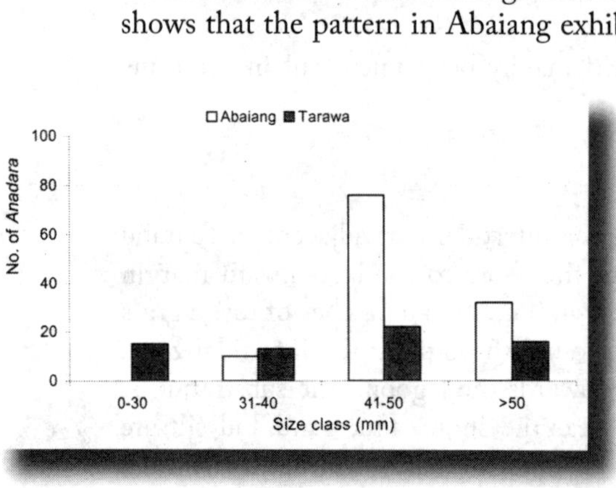

Graph 1. Size-class distribution of Anadara for Abaiang (N=118) and Tarawa (N=66) study sites.

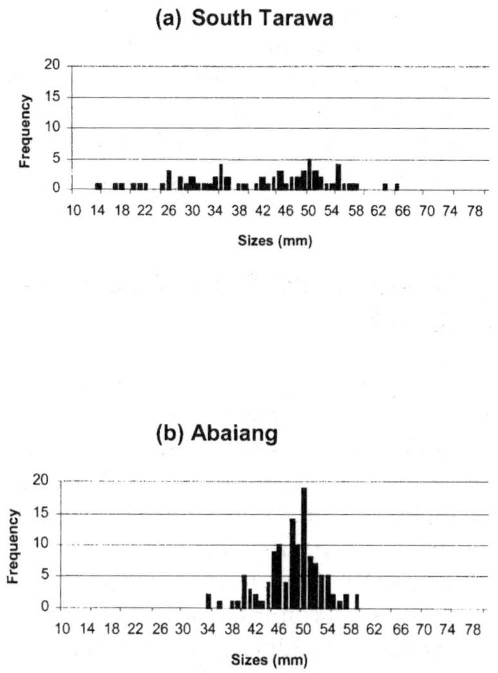

Graph 2. Size frequency histogram of Anadara for the study sites: (a) Tarawa (N = 66), b) Abaiang (N= 118)

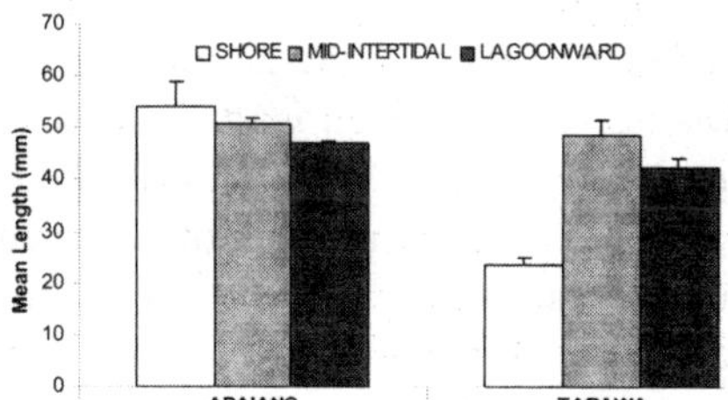

	ABAIANG	TARAWA
SHORE	54.05	23.50
MID-INTERTIDAL	50.73	48.48
LAGOONWARD	47.23	42.55

Graph 3. The change in average length of Anadara, moving from the shore through the intertidal to the lagoonward area of each site.

patterns for two sites you find peaks occurring at different shell lengths, this suggests differences in the timing of recruitment. The absence of a peak may also signal that any clams that did recruit into the population at a particular time did not survive long.

The distribution pattern moving from the shore towards the lagoon also varied between the sites (Graph 3). In Abaiang, the largest shells were found closest to shore and shell size decreased as one moved out into the lagoon. In Tarawa, shells found close to shore were on average much smaller than those farther out. As in Abaiang, there was evidence of a slight decrease in size moving from the mid-intertidal to the lagoonward edge of the flats.

WHY ARE THE *ANADARA* POPULATIONS IN TARAWA AND ABAIANG DIFFERENT?

Abaiang had a significantly higher density of *Anadara* compared to Tarawa. Given that the two sites were very similar in many physical features but experienced very different harvesting pressures, one could conclude that harvesting pressure has had an impact on the density of clam populations on Tarawa. Simple differences in density may, however be misleading. In 1999, the mean density of *Anadara* in South Tarawa was estimated to be 2.9 $\pm$ 0.7 m^{-2}. In 1995, Paulay (1995) recorded an average of 4.2 $\pm$ 0.8 *Anadara* m^{-2}. Yamaguchi et al (1993) documented an even lower density of 0.9 $\pm$ 0.7m^{-2} several years earlier. These differences over time might reflect variations in fishing pressure, but they could also be attributed to fluctuations in recruitment patterns or even the effects of catastrophic events such as cyclones.

Variations in density that occur between sites at one point in time could be related to different histories, for example, differential timing and success of recruitment. There may also be physical and biological factors that affect *Anadara* and vary from place to place, even within study sites. These factors include the density of turtle grass, the amount of coarse sand or silt on the bottom, degrees of pollution from land-based sources and presence of animals that might compete with clams for space and food. Compounding such factors are human activities including trampling of turtle grass, harvesting of other animals and pollution. These physical, biological and human factors are difficult to measure and their combined effects can have various impacts on *Anadara*. In this case, variations within study sites were not statistically significant. Although potentially important physical and biological features were not measured, the two sites did appear to be reasonably similar with the exception of organic pollution, which is undoubtedly greater on Tarawa.

The second significant difference between Abaiang and Tarawa sites was in the size of *Anadara*, which was relatively smaller in the heavily harvested site. In the less harvested site, clams closest to shore were the largest. This may reflect the natural condition for this species, as a similar trend was observed in unharvested areas in Fiji (Maybin 1989, Butler 1983). In contrast, clams were particularly small in inshore areas of Tarawa. This could logically be the result of intense harvesting pressure in this part of the sand flats.

Previous research showed that removal of many large shellfish causes a reduction in abundance as well as in the mean size of shells (Sharpe and Keough 1998). Casual observation of the shellfish market reveals that large *Anadara* are particularly targeted for sale. These 50-60 mm shells are the part of the population that is missing from Tarawa sand flats.

Recruitment peaks detectable in the Abaiang population were flattened out in the Tarawa population. This could be linked to harvesting pressure that is sufficient to fish out pulses of clams as quickly as they recruit into the population.

The smallest size class of shells was missing from Abaiang samples. One possible explanation is that recruitment failed in the recent past. Alternatively, juveniles may settle further offshore in the dense turtle grass beds and then migrate towards shore. Further research is required on this topic.

Judging from the evidence presented above, we conclude that fishing pressure is a very likely contributor to the differences seen in the *Anadara* populations of Tarawa and Abaiang. Further research into other potential contributing factors is required.

THE SOCIAL SIGNIFICANCE OF *ANADARA*

Women play a very important role in the *Anadara* fishery, and have a growing dependence on the resource for subsistence purposes and income generation (Tebano, pers comm). Seafood is the main component of most Pacific Islanders' diets and thus, subsistence fishing is very important to local culture.

In Kiribati, despite a growing dependence on food imports, fishing is still important. Fish and shellfish consumption remains relatively high. Mees et al (1988) reported that the average shellfish catch per person per day (presumably for local consumption) is approximately 1.3 kg. There is a marked difference in fish consumption between South Tarawa and the outer islands. Fish consumption is considerably lower in South Tarawa (320 grams/person/day) compared to outer islands (565 grams/person/day) (Zann 1985w), which suggests that local shellfish is a critical source of protein.

Modernization and monetization have created the need for money in urban and rural areas. Urban consumers may have no option but to purchase fish and shellfish as well as other foods. To meet their need for cash, women, who are primary providers for the household, try to supplement household income in whatever ways they can. Unemployed women living in crowded households are likely to enter into semi-commercial harvesting of marine resources. The *Anadara* fishery is attractive to such women because the shellfish lie close to shore and no special equipment is needed to harvest them.

The sale of *Anadara* contributes significantly to household incomes in Kiribati (Fay-Sauni 2001). Studies have shown that the involvement of women in fisheries activities contributes to the improvement of the socio-economic status of households (Steele 1993; Chapman 1987; Lal and Slatter 1982). When women earn income and contribute financially in this way, they also begin to participate more actively in household decision-making.

If the relatively smaller size and density of *Anadara* in Tarawa (compared to Abaiang) is linked to fishing pressure, this is likely a consequence of an increase in the numbers of *Anadara* fishers and the commercialization of the resource.

Harvesters with shellfish in Tarawa Lagoon.

The consequences of overexploiting *Anadara* would greatly affect the fishers who rely on the resource. The Canada International Development Agency (1993) found that when a resource is overfished, women may be displaced to more menial fisheries tasks. Their traditional roles and responsibilities may then be taken over by men, at which point women's input and influence in community decision-making are reduced. In the context of the family, the loss of income and decrease in responsibility for financial management of the household leaves a woman and her children more dependent on the decisions of the male partner. In the case of female-headed households, loss of the woman's normal income can jeopardize family survival. Also, the woman and the community as a whole may lose technical knowledge.

There are few well-paid employment opportunities for Pacific Island women who have limited education. There may be few income-earning alternatives of any kind in specialized or remote communities. Women's core experiences lie within the household and economic avenues that use this expertise and experience are rare. If women are displaced from fishing, the knowledge and skills developed in the context of fishing activities are not easily transferred to other occupations. Moreover, their displacement from local fisheries activities represents a loss to the industry as a whole, and to the local economy, since the human resource potential of the community is not fully utilized. It is therefore important that the marine resources that support women's fisheries are monitored and managed to avoid serious problems.

Management Implications

As noted above, there is no consistent trend of decline in *Anadara* density on Tarawa when one compares current data with previous studies. On the other hand, differences between Tarawa and Abaiang suggest that at least the shallow waters of Tarawa lagoon are under serious pressure. The harvesting of *Anadara* may, nevertheless, be at a level that will not adversely affect future reproduction and recruitment.

This study did not establish the maximum sustainable yield for the fishery. Nevertheless it is suggested that as a precaution, to reduce the risk of depleting the reproductive *Anadara* stock, management is needed.

Various studies have shown that fisheries ought to be managed in a way that maintains the stock biomass at a level that is capable of producing maximum sustainable yield on a continuing basis (King 1995; Munro and Fakahau 1993). King (1995) states that this is a conservative or precautionary approach to management that is likely to become the standard for all fisheries management plans. Management of peri-urban intertidal areas is often necessary. For example, in Sydney Australia, a limit is imposed on the number of shellfish that can be collected. Only 50 bivalves (*Anadara* sp. included) may be taken at one time, with fines up to AUS $5000 for illegal collecting (NSW Fisheries 2001).

Although various management measures could be imposed on the Tarawa shellfishery, the effect of regulation on the welfare of the fishers needs to be considered. Most important is the impact on fishers who rely totally on the resource. Measures such as size limits and catch quotas could be effective, but the costs of enforcement and, most importantly, the adverse effects at the household level, could be prohibitive. Rotational closures and Marine Protected Areas (MPAs) have proven effective in Fiji and other countries. For example, in Ucunivanua village in Verata, Fiji, an MPA for *Anadara* has resulted in the increased density of smaller size classes of *Anadara* (A. Tawake pers comm). This signals the successful recruitment of shellfish that will be allowed to grow, reproduce and enhance surrounding fishing grounds.

Closing offshore *Anadara* fishing areas could also be effective in restoring intertidal stocks. Tebano and Paulay (1995) found that Tarawa lagoon has various shoals that support lush seagrass beds with a high density of *Anadara*. These are currently being targeted by male divers (pers obs) but if protected from harvesting, they could provide larvae to boost recruitment into the intertidal fishery.

In most cases, minimum size limits are used to alleviate growth overfishing of invertebrates. The appropriate limit depends on growth and mortality rates of the shellfish in question. Published literature shows that *Anadara* in Kiribati reaches maturity at sizes of 38 mm (females) and 42 mm (males) with partial spawnings coinciding with the full moon (Tebano 1990, Tebano in prep). The growth rate of *Anadara* is 1.66 mm per month (Tebano pers comm). Tebano recommends that the minimum size limit for South Tarawa be set at around 42 mm.

Social Considerations in Devising a Management Strategy

The problem that can be envisaged with respect to imposing minimum size limits is that it is difficult for fishers to know the exact length of a shell in millimetres (pers obs). According to Tebano (pers comm), this was the problem when the fisheries department tried to manage other resources using size limits. A size limit should therefore be related to the size of the harvester's thumb or little finger.

In the case of bonefish, *Albula vulpes,* the Fisheries Department decided to impose size limits in conjunction with closed seasons. Fishing was closed during the spawning run of the bonefish, which is seven days before and after the full moon. This scheme was implemented through a collaborative effort between fishers, the Department of Fisheries and the local councils (Teinainano Urban Council, Betio Town Council and North Tarawa Council). It worked because it was the fishers' initiative that was supported through the elders of the village *maneaba*[1] system. Urban town councils need to be involved because they have the power to enforce any management decision through bylaws (T. Tebano and A. Tawake pers comm).

A similar approach can be taken with regard to the *Anadara* fishery in South Tarawa. A proposal could be made for a minimum size limit of 42 mm in conjunction with a closed season during the clams' spawning period, which is seven days before and after the full moon (Tebano pers comm). A collaborative effort involving the Department of Fisheries, other government departments, urban town councils and the traditional owners of the lagoon, together with the fishers, is needed to implement this. Contact with female fishers in South Tarawa could be made during fund raising activities because women often attend these events (pers obs).

Legislation and bylaws should be the end product of a grassroots awareness programme. Awareness can be created through radio and newspaper advertisements as well as through informal gatherings. Public notices in both local and English languages should be posted in areas frequented by fishers, for example near markets, big stores and public areas. Suitable sites in South Tarawa include Korobu-Banraeaba (opposite the Adventist church building), Teaoraereke (former water reservoir area), Eita (outside Aontoa store) and Bikenibeu (in the old hospital area).

Lack of awareness of the adverse consequences of overexploitation has hindered the participation of women in management planning. Because women are the main users of nearshore marine resources, they should be made aware of the consequences of the unwise use of resources, and the benefits of sustainable harvesting.

Women's Role in Resource Management

In the past, fisheries managers carried out research as a means of monitoring the status of the resource but often did not incorporate the views of fishers. As a result of misinformation and the non-participation of fishers in information sharing, management and decision-making, fishers were not encouraged to employ or adopt sustainable harvesting tactics. Participation of women fishers at the grassroots level (in workshops or community consultations related to resource management and conservation) is therefore very important. Women need to be involved in planning, devising resource management plans and implementing activities. Their array of skills and expertise could be significant assets for the management of various marine resources.

STRATEGIES FOR SUSTAINABILITY

This case study concludes that there is a significant difference in density and size structure of *Anadara* between harvested and unharvested *Anadara* beds in Kiribati. This difference can be attributed to the greater fishing pressure in the urban area, and also to the preference of fishers for harvesting the larger shells and for working in the most easily reached inshore areas. Although fishing pressure is an obvious factor, other influences such as habitat differences, recruitment variations, sediment size and composition, and the percentage cover of seagrass, cannot be ruled out. More detailed research would be needed to determine whether and how these factors affect this resource.

In light of the major findings the following recommendations are made:

- More research is needed on the factors affecting density and size such as habitat differences, recruitment variations and environmental changes.
- There is a need for continuous monitoring of the *Anadara* catch rates, both in shallow and deep waters, to help assess the effects of harvesting.
- More research on reproduction and growth of *Anadara* is required to be able to determine the appropriate size limits for management purposes.
- Women must be encouraged to become involved in management and conservation of the *Anadara* resource, especially in the planning and implementation of appropriate management approaches. Women fishers can be reached through various local networks such as community and church groups. They can be encouraged and enabled to engage in resource management activities if these are properly designed to be participatory and accessible to women.

REFERENCES

Butler, A J. 1983. *A Preliminary Examination of Populations of the Kai-koso, Anadara cornea (Reeve) Near Suva, Fiji.* A report to the Institute of Marine Resources, Suva: University of the South Pacific, 28p.

Canadian International Development Agency 1993. *Women and Fisheries Development.* CIDA, Canada.

Chapman, M. D. 1987. Women's fishing in Oceania. *Human Ecology*, Vol. 15(3):267-288.

Chen C. W., D. K. Leva & W. J. Kimmerer 1995. Water chemistry and lower levels. In: Abbott, R. R. & J. Garcia (eds) *Management Plan for Tarawa Lagoon, Republic of Kiribati - Management Plan.* Santa Cruz: Biosystems Analysis Inc. 3:1-34.

Douglas, N. and N. Douglas (eds) 1994. *Pacific Islands Yearbook*, 17th edition, Fiji Times Ltd., Suva, Fiji.

Fay-Sauni, L. 2001. *The Involvement of Women in the Anadara (Gray, 1847) Fishery in South Tarawa in Kiribati and Suva Peninsula in Fiji.* MSc Thesis, University of the South Pacific, Suva, Fiji. 126p.

Kimmerer, W.J. & T.W. Walsh, 1981. Tarawa Atoll lagoon: Circulation, nutrient fluxes and the impact of human waste. *Micronesica* 17:161-179.

King, M.G. 1995. *Fisheries Biology and Management*. United Kingdom: Fishing News Books, Blackwell Science, p.161-164.

Kiribati Government 1996. *1995 Kiribati Census Report*. Kiribati: Ministry of Finance and Economic Planning.

Kiribati Government 2002. *Kiribati Statistical Yearbook 2002*. Kiribati: National Statistics Office, Ministry of Finance.

Lal, P. N. & C. Slatter 1982. *ESCAP/FAO Project on Improving the Socio-economic Condition of Women in Fisheries: Relevance of the Fiji Study to Other South Pacific Countries.* Working Paper for Meeting on Fisheries, South Pacific Commission, Noumea, New Caledonia, 4p.

Maybin, J. A. 1989. *Ecological and Taxonomic Aspects of Anadara (Mollusca: Bivalvia) in Fiji and Some Neighbouring Islands Groups.* MSc thesis, University of the South Pacific, Fiji, 88p.

Mees, C. C., B. M. Yeeting & T. Taniera 1988. *Small Scale Fisheries in the Gilbert Group of the Republic of Kiribati.* Government of Kiribati, Ministry of Natural Resource Development, Fisheries Division.

Munro, J. L. & S. T. Fakahau. 1993. Appraisal, assessment and monitoring of small scale coastal fisheries in the South Pacific Region. In A. Wright & L. Hill (eds), *Nearshore Marine Resources of the South Pacific*, ICOD, Canada, p.15-72.

NSW Fisheries 2001. NSW Fisheries 2001.
http://www.fisheries.nsw.gov.au/conservation/psa.htm

Paulay, G. 1995. Benthic ecology and biota of Tarawa Atoll lagoon: Influence of equatorial upwelling, circulation, and human predation. In R. R. Abbott & J. Garcia (eds), *Management Plan for Tarawa Lagoon, Republic of Kiribati - Management Plan*. Santa Cruz: Biosystems Analysis Inc. 3:1-34.

Sharpe, A. K. & M. J. Keough 1998. An investigation of the indirect effects of intertidal shellfish collection. *Journal of Exp. Marine Biology and Ecology* 223:19-38.

Steele, S. 1993. Literature summary: Women in fisheries development. In *Workshop on People, Society and Pacific Islands Fisheries Development and Management: Selected Papers*. South Pacific Commission, p.31-37.

Tebano, T. 1990. *Some Aspects of Population Biology and Ecology of the Cockle, Anadara maculosa in the Gilbert Islands, Republic of Kiribati.* MSc thesis, Faculty of Science, James Cook University, Australia.

Tebano, T. (in prep). *The Biology of Te Bun Anadara [Gray, 1847] (Bivalvia: Arcidae).* PhD thesis, School of Pure and Applied Science, University of the South Pacific, Suva, Fiji.

Tebano, T. & G. Paulay 1995. Variable recruitment and changing environments create a fluctuating resource: The biology of *Anadara* (Bivalvia: Arcidae) in Tarawa. In R. R. Abbott & J. Garcia (eds) *Management plan for Tarawa Lagoon, Republic of Kiribati - Management Plan*. Santa Cruz: Biosystems Analysis Inc. 3:1-34.

Vunisea, A. 1997. Women's fishing participation in Fiji. *SPC Women in Fisheries Information Bulletin* #1, October 97, SPC, Noumea.

Yamagouchi, M., B.Yeeting, K.Toyama, T.Tekinaiti & T. Beiatau 1993. *Report on a Preliminary Study on Molluscan Resources in Tarawa Lagoon, with Special Reference to 'te bun' (Anadara maculosa).* Fisheries Division, Ministry of Natural Resources, Republic of Kiribati.

Zann, L. 1985. Traditional management and conservation of fisheries in Kiribati and Tuvalu atolls. In K. Ruddle & R.E. Johannes (eds), *Traditional Marine Resource Management in the Pacific Basin: An Anthology*. Jakarta: UNESCO/ROSTSEA: p.53-78.

NOTE

1. The *maneaba* is a traditional meeting house for village elders in Kiribati

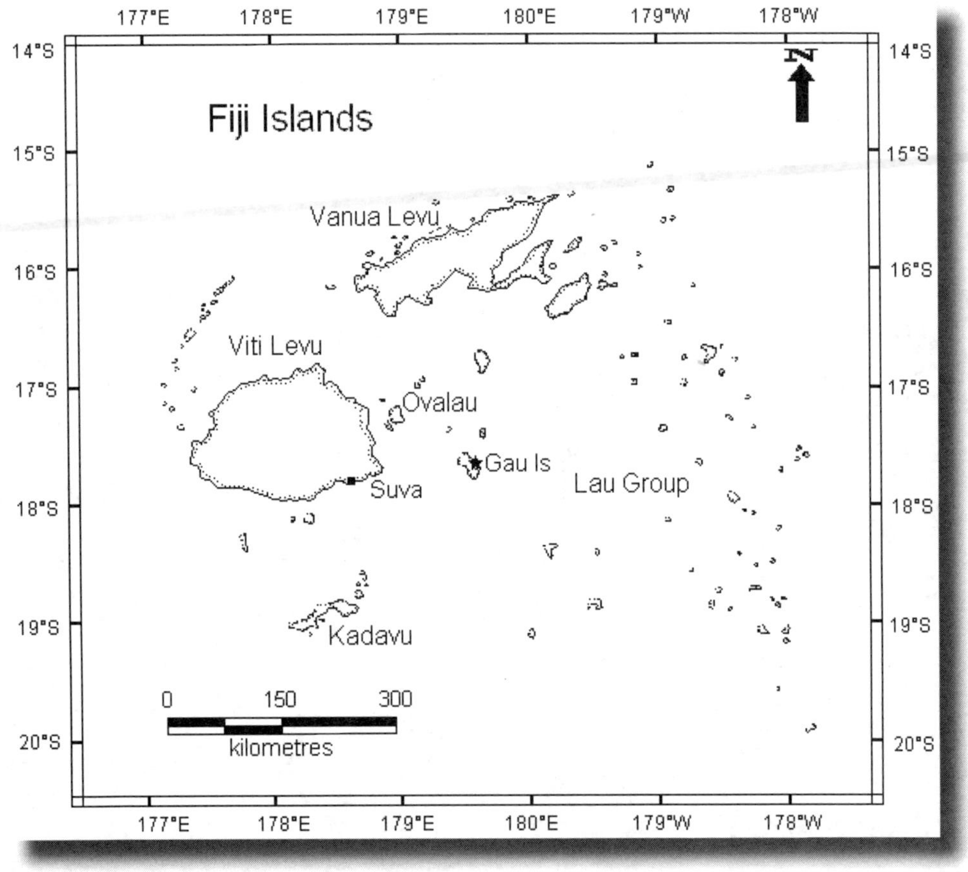

Map 1. Map of Fiji showing the location of Gau Island.

6 *Yavi Rau:* The Customary Fish Drive In Malawai & Lamiti Villages, Gau Island, Fiji

Joeli Veitayaki and Waisake Paul Vesi

Yavi rau, the customary fish drive, is a communal event in Gau, Fiji that is organised only occasionally. *Yavi rau,* which means hauling the *rau,* the vines wrapped in coconut fronds, is most commonly organised during the festive season or on special occasions. In Malawai and Lamiti Villages on the eastern side of Gau Island, the fish drive is conducted in the same way because the people are closely related and the reefs of the two villages have similar features. Other villages have different marine environments and therefore have slightly varied styles of *yavi rau.* In this photo essay, we share a glimpse of a traditional fishing expedition that has survived into modern times. The essay provides a description of the process and comments on some of the features of this cultural event.

Once the decision to have the *yavi rau* is made by the village council, a presentation is made to the heads of the land owning units (*mataqali*) to seek permission for the use of their land, fishing ground and coconut leaves. This tradition is associated with the Fijian practice that customary fishing areas are owned by different, but closely related, social groups that regulate their use and exploitation. People are expected to use their own customary fishing areas, and those seeking to use grounds belonging to others are expected to get permission from the owners. Only then can the preparations begin. The initial activity is the annual clearing of the village burial grounds. It is believed that the fish drive will be successful if the burial sites are cleared. Preparations related directly to the fishing are laborious and take up to two days.

Over several days, the men gather vines (*walai*) from the forest and coconut leaves from nearby plantations. The vines must be of the correct size so that, on one hand, they do not break easily when stretched but on the other hand, they are not too thick and heavy. The vines are straightened by pulling, stretching and hanging them from trees along the beach.

Coconut fronds are neatly wrapped around the straightened vines to make the *rau.* In this activity, women and children join the men folk. Normally, the villagers make *rau* that is four to six kilometers in length, which is sufficient to cover the fishing ground.

On the designated fishing day, *rau* is loaded into punts to take it to the reef. The use of motorized fiberglass punts has made the placement of vines a lot easier. In the past, this was a physically demanding exercise, as men used to pole the punts. Today, two or three punts are all that is needed.

A male member of one of the fishers' families or someone that is familiar with fishing in the village normally commands the *yavi rau.* During

Preparing the rau vines that will be joined to make long ropes.

Boys weaving coconut fronds around the vines to make the rau.

The stretched vines will be wrapped with coconut fronds to make the rau

the drive, the commander is the sole authority. He determines the sequence of events during the *yavi rau* and is responsible for the success of the fishing expedition as well as the safety of all the villagers involved in the fishing. In some instances, verbal and physical abuse is meted out to ensure that the objectives of the fishing are achieved.

Weaving the coconut fronds to vines to make the rau.

On the fishing day, the *rau* is taken out to the fishing ground at the first sign that the tide is ebbing. The preference is for an ebbing tide that occurs early in the morning. This is because the fish are less likely to have been disturbed in the night and consequently would have come close to the shore and would still be there when the *rau* is placed.

Large numbers of people are required so that the *rau* is adequately manned and properly handled as the people are marshalled ashore. All the able-bodied men and women are expected to be at the fish drive. Children are encouraged to join their parents on this occasion while neighbouring villagers, particularly the women, are also welcomed. However, the women are always advised to declare their

It takes many hands to deploy the rau and pull it to shore.

pregnancy and not to join the fishing if they are expecting. Fijians have a belief that the involvement of pregnant women will result in little catch for everyone.

The fish drive is based on the principle that *rau* in the water is an impregnable barrier that can be used to chase fish. Under the direction of the master fisher, the people haul the *rau*, as they slowly walk towards the shore. The *rau*, which is placed along the inner edges of the reef, is transformed as it is hauled landward. From its initial linear shape, the *rau* becomes a semi circle as the two ends are drawn together. The circumference of the semi circle that is formed is continually reduced in size as the people move landward. The fishing party proceeds at a rate that will ensure that the *rau* reaches the sandflat at low tide; this is where the fishing will conclude.

By low tide, the fishing area encircled by the *rau* has been reduced to a fraction of its original size and has become a full circle. Excess *rau* is discarded on the drying sandflat and used to reinforce the circle by doubling and tripling its thickness. The circled areas are continually reduced until small enough to allow for efficient gathering of the catch. By this time, the fish are literally on top of one other because of their concentration in the small area and the low tide.

The men, women and children wait for the order to start gathering the catch. Women use scoop nets to gather smaller fish while the men use knives and spears to catch the larger fish and any fish such as unicorn fish, garfish and barracuda that can cause injuries.

The catch is gathered and taken to the village where it is shared among all the people who were involved in the fishing and in the preparations. Prior to dividing up the catch, most of the women and children would have taken a portion of the catch for their own lunch. Nearly all of the villagers would have fish for lunch on this day even if they did not receive a formal share of the catch.

Under the command of the master fisher(s), the rau is systematically put in place as the tide is receding

Fish of all kinds are speared in the shallow water surrounded by the rau.

Fish is harvested and placed in the punts.

The fish is shared to all the people involved in the preparation and fishing.

The fish catches are so large that people are expected to prepare some portion of the catch for storage. Women and children are involved in the chores associated with scaling, gutting and preservation. Fish is fried, smoked or cooked in a salty solution so that it is preserved for use in the following days. Smoked fish keeps the longest and can be used a week or so after it was smoked. In some instances, the smoked fish is reheated after every two to three days.

Yavi rau is a traditional fishing method that is still used today and is likely to be used in the future. The involvement of young people is a sign that knowledge is being transferred to the next generation. The fishing method is particularly efficient and is used when big catches are needed.

Fish drives are also destructive. People create as much disturbance as they can while they swim and walk towards the shore. The *rau* commonly gets caught on rocks and coral, which are broken. People trample on the coral and other substrates and collect all of the resources they see as they move landward. On land, the coconut trees are stripped of their leaves. In a way, the irregularity of fish drives moderates the level of destruction. Yet for that short time when it is being planned and executed, all the attention is focused on securing the maximum amount of fish for the occasion.

The fish drive is a communal activity that demonstrates an intimate knowledge of the fish. The fish are herded in by the *rau*, which must cover a large area to be effective. The fishing is organised according to customary practices and there are particular roles for the different social groups in the village. The catch is shared among all people who participate in some way in the fishing. The owners of the fishing grounds and coconut plantations receive shares of fish, as do the owners of boats and people from outside who worked in the preparation of the *rau*.

At the end of the day, the village men gather around the kava bowl to acknowledge the day's effort, to share stories of the fishing and the special events of the day, and to plan better fishing on the next occasion.

Personally, we were happy to have shared this experience with our family members, all of whom were unfamiliar with fish drives. The young children were very excited and we were worried that they might overstep their social position. In the village, fish belong to the warriors (*bati*) and we were not.

We were told that the catch was good and that this success must be due to the marine resources management now undertaken by the people. It was particularly pleasing to hear. It is possible to deduce that the resource management rules will not be relaxed soon because they are working well and are providing good catches now and in the future.

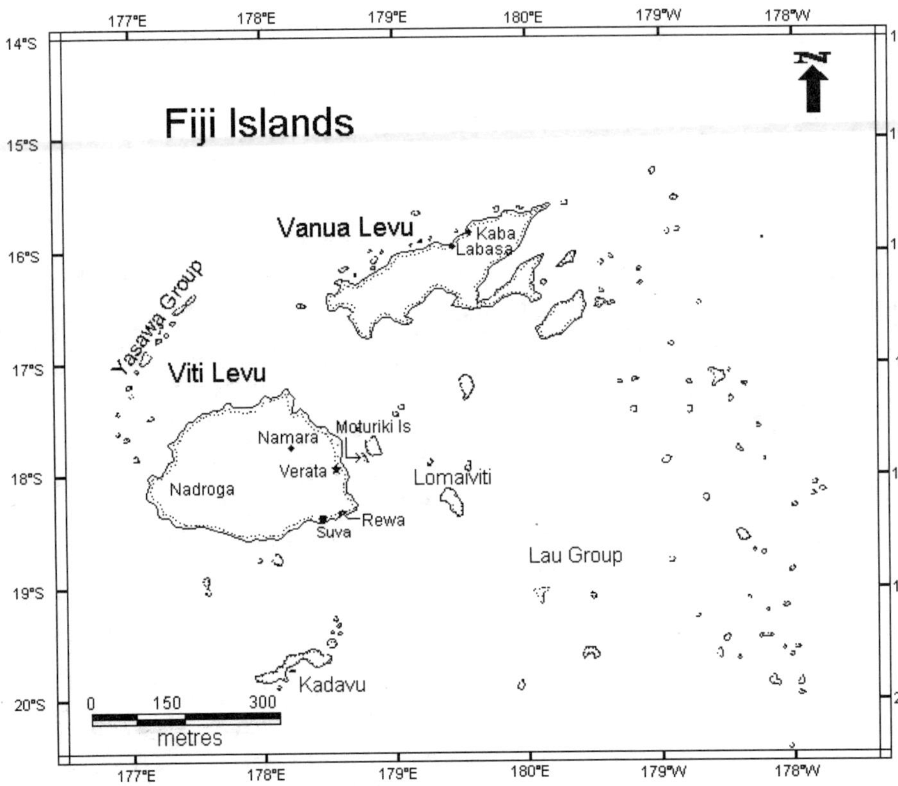

Map 1. Map of Fiji showing the various coastal areas where information was collected.

7 Women's Changing Roles In The Subsistence Fishing Sector In Fiji

Aliti Vunisea

"If I was not out here fishing, what else would I be doing? I look forward to these fishing trips. This is my only chance to get out of the house and join other women from the village. I can test my skills at line fishing, crab fishing, shells collecting and other such activities. I can also catch up with gossip and news from other women. I can buy food from the shop but the feeling of a fish pulling on my line is more exciting"

— Timaleti Yavu, a fisherwomen from Navakasiga in Bua

GENDER ROLES IN SUBSISTENCE FISHERIES OF THE PACIFIC

Subsistence fishing in Fiji - fishing for food and to support one's basic livelihood - is often dominated by rural women. Both men and women actively participate in a wide range of fishing activities, but there are defined gender roles. Women predominate in the subsistence fisheries while men dominate the industrial and artisanal sectors. Gender roles within different small-scale fisheries are not always the same, nor are they static. Women often sell seafood but in some cases they may not participate in the harvest at all. In other cases, for example the *kai* (freshwater mussel) fishery in Fiji, sellers are sometimes men but the harvesters are women.

The subsistence fishery as defined in this case study refers to all activities related to fishing, e.g. pre-harvest, harvest, post harvest processing and distribution. Subsistence fishing activities are associated with generations of fishing lore, skills and long-standing, traditional institutions. Subsistence fishing not only provides a basic source of food and income to rural Fijians but also to urban and peri-urban populations who depend on catches from semi-subsistence fishing for their basic seafood needs. In recent decades, fisheries have become increasingly commercialized, making it more difficult to distinguish between artisanal and subsistence fisheries.

Most studies of subsistence fisheries that have been done in the Pacific do not record women's participation or, if they mention women's involvement, they do not attempt a detailed documentation of their fishing activities, or of the knowledge and skills that fisherwomen possess. Schoeffel (1984), Talagi (1989) and Matthews (1993) acknowledged women's substantial participation and the need for a thorough assessment of this participation. In Fiji, Lal and Slatter (1982), Ram-Bidesi (1993), Slatter (1995), Vunisea

(1994, 1995), Robinson et al (1996) and Veitayaki (2000) all acknowledged women's involvement and the lack of information on their many fishing activities. All writers refer to the immense contributions of women fishers but there is little mention of the value of these fishing activities to the economic, cultural and social well being of households and communities.

In this case study I will explore how gender roles in fishing have changed through time, as women have increasingly taken up fishing and become involved in related activities such as distribution and sales. Through their increasing participation in fisheries, women have acquired monetary and social empowerment. I will discuss how women have employed coping strategies to support their fishing activities and as a result, have carved niches for themselves in the fisheries development process. For example, women in rural villages have established networks that help them to adapt to changes brought about by the modern market economy. Finally, I will point out some ways in which the efforts of women in fisheries can be supported and their working conditions improved.

Women sell garden and fishery products on the sidewalks near the main market in downtown Suva, Fiji.

CASE STUDY METHODS

Most of the information in this paper was collected as part of Master's thesis research in 1995-1996. After a preliminary reconnaissance of the study area, research was conducted in the villages of Ucunivanua and Kumi in the Verata district of Tailevu Province, Fiji (Map 2). Fieldwork took place over a four month period, during which time I stayed in Ucunivanua and Kumi Villages for periods of up to a week. Most of the information collected has been updated since the original research was undertaken, most recently through field studies conducted in January 2004.

The methods used for the original survey consisted of the following activities: literature and archival research; formal approaches through the Fijian District fisheries officers and civil servants in the study locations; questionnaire surveys conducted with one woman from each of 10 sample households in each village; group discussions and in-depth interviews with men and women identified as having specific fishing knowledge; participatory observation; creel inventories; a market survey; and purposeful sampling of fishers and other households not included in the initial random sample.

Comparisons with and references to other areas of Fiji are derived from other fieldwork undertaken from 1996 to 2002, in the course of NGO and USP projects related to marine resource use and management. This work involved conducting provincial, district and village based workshops,

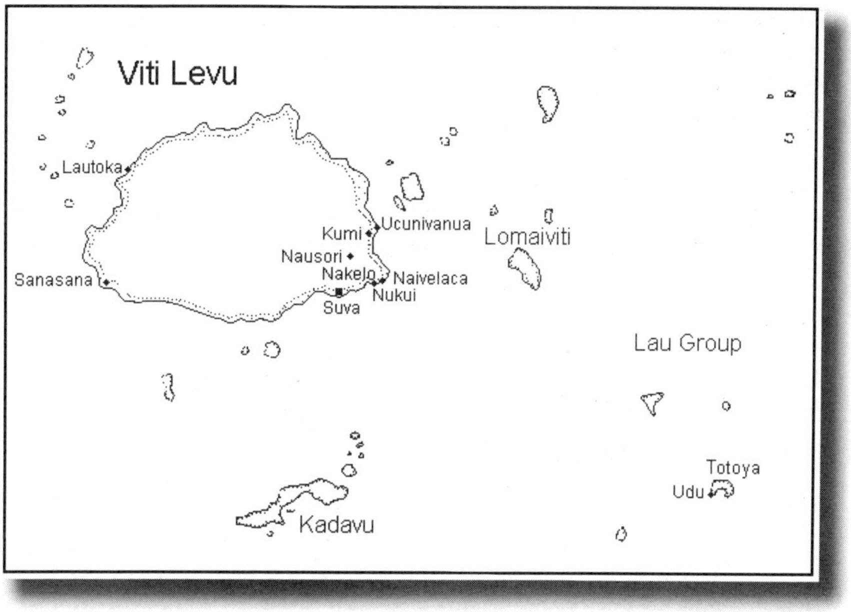

Map 2. Viti Levu, showing villages of Ucunivanua, Kumi, Totoya, Lautoka, Suva, Udu, Naivelaca, Nukui, Nausori, Nakelo, Sanasana

working at the community level and interacting with resource users, especially women. I had the opportunity to observe the social dynamics of village fisheries and the similarities and differences that exist among coastal communities in Fiji.

MARINE RESOURCES IN VERATA

Of the seven villages within the Verata District, all but one are coastal. Verata is blessed with easy access to extensive foreshore flats, sheltered offshore reefs and diverse offshore fishing locations. During very low tides the intertidal area and associated reefs are often exposed and this is where fishing for subsistence purposes is concentrated.

1) The main fishing zones in Verata include:
2) the on-shore area *(baravi)*
3) the intertidal sand and mud-flat zone *(dela ni mati)*
4) the inshore reef area *(cakau)*
5) patch reefs *(cakau yawa)*
6) distant barrier reefs *(cakau sese)* and
7) the deep open ocean and more distant reefs *(waitui loa or wasawasa).*

TRADITIONAL FISHERIES KNOWLEDGE

Generations of Fijian fishing people have perfected fishing techniques and acquired an intimate familiarity with fish habits and behavior. Some of this local knowledge is shared even by the young children in coastal villages. People with in-depth fisheries knowledge and skills are known and respected in their communities. Despite the existence and acknowledgement of the *gonedau* or traditional master fishermen clan, highly skilled fishermen and women can be found in any family.

Women as well as men hold important ecological knowledge. The wealth of knowledge people have of their fishing areas is shown in their understanding of fish behavior and of the importance of tides, lunar cycles, winds, weather and the seasons. The effects of tides on marine life are well known by women and this knowledge helps them to select target species and locations. For example, certain fish species migrate through the channels in reef flats at particular stages of the tide. Women therefore fish with hook and line along reef edges on the incoming tide and over reef flats at high tide. At the turn of the tide, fishing is concentrated in outer lagoon areas. As well as the daily rhythm of the tides, the timing of exceptionally high and low tides (spring tides) as well as wind direction and intensity are important influences on the success of fishing.

Changes in the weather also affect fishing strategies. For example, gleaning is actively pursued soon after a storm has passed by. In the Rewa delta area where I am from, people come out in force looking for mud lobsters (*Thalassa anomola* or *mana*) and crabs after such storms. Fish feeding near the shore may panic at the sudden onset of a storm and rush for shelter, sometimes ending up on dry sand. At Naivilaca in November 1994, a sudden rainstorm left schools of snappers, *kabatia ni cakau* (variegated emperor), *kawakawa* (marbled cod), *ta* (unicorn fish) and *ulavi* (blue-barred orange parrotfish) stranded on the sandflats. People deduced that these fish were probably feeding over the sandflats with the in-coming tides when the storm struck, and the clap of thunder and sudden change in wind direction left them frightened, disoriented and helpless.

Thunder and lightning are considered to cause many species to leave their hiding places. People use this knowledge to lure different species into the open by "creating thunder" *(kurukuru)*. This technique is widely used during fish drives in Ucunivanua, Kumi, Nukui, Totoya and in other coastal locations in Fiji (Map 2). People bang rocks together under water to frighten fish into waiting nets or traps. When women dive for sandfish or dairo in Nukui, paddles or oars are used to beat on the surface of the water before fishing commences. This is done so that the animals will come out from where they lie partly buried in sand.

Lunar cycles also affect the success of fishing for different species. During new moons *(vula cabe)* there are aggregations of certain reef fish species and some snappers such as kawakawa are regarded as being unusually docile. In Sanasana in the Nadroga province, women related how they fish in groups over these fish aggregations. The women count on this fishing opportunity to provide excess fish that can be sent to relatives in town.

EXPERT FISHERWOMEN

It is commonly assumed that women's gleaning activities involve a simple process of gathering resources that are lying on the reef flats. In fact, their activities are varied and involve the intricate use of hands, feet and eyes as well as a range of tools. To succeed, gleaners also require an in-depth knowledge of the habitats of species, inter-species relationships and the seasonal and temporal patterns of abundance. Women are more aware than men of the minute details of the inshore reef flats, the different names given to target species, and their behaviour. Skills such as those used to gently pry shellfish from under rocks and to clamp bare hands over crabs and fish give women a unique edge over men when fishing in shallow waters.

Octopus caught by expert fisherwomen are often smoked for sale in the city.

In Totoya and in Ucunivanua, when women took me fishing they would take great pains to explain how to identify fish holes, find shellfish buried under sand, identify crab tracks and follow octopus trails. I assured them I knew what to look for. Later, after my long, fruitless search, one of the women walked up and poked at an octopus in a lair right next to where I was standing. After all my training and fishing skills accumulated through the years and despite all the assistance, descriptions, prodding and hints, I failed to find an occupied octopus lair! Meanwhile, my companions were overladen with octopus, which they slung around their shoulders.

One of the older women in the group, trying to make light of my situation, said that I should not worry about not finding an octopus. She kindly pointed out that after years of practice, the local women know where the lairs are, how the octopus feed and where they go. I was convinced that women are the experts in this type of fishing.

Seasons are correlated with the availability of some marine products. For example, the cold season from the end of June to August signals the abundance of octopus and seaweed. Land plants may also serve as markers for the availability of certain marine species. The octopus and seaweed season, for example, correlates with the flowering of dadap or drala *(Erythrina variegata var. orientalis).* Many seasons are centred on yam planting and harvesting, presumably because of the importance accorded to yams in traditional Fijian societies. When yams begin to grow *(vula i tubutubu)* seaweeds also grow and these disappear when the yams die. Generally, good harvests of yams are indicative of good harvests of seaweed. Such associations are interpreted as a result of "the hand of God". In fact, some of the women I questioned referred to the sea as *i teitei ni kalou,* God's garden.

As in the past, traditional fishing activities of both men and women are to an extent influenced by people's belief in the supernatural. The supernatural is both respected and feared by people and this means that people have reservations about breaking certain taboos or fishing rules. Fishing people fear that misfortune or sickness will befall family members or that there will be mishaps at sea as a result of some misdeed.

FISHING METHODS

The traditional fishing methods employed by both men and women can be classified into seven major categories: gleaning; net fishing; line fishing; use of set and mobile barriers; night fishing; fish poisoning and spear fishing. The major occupation of women, both historically and in modern times, is gleaning or *vakacakau.* In the discussion below, the various activities that women fishers engage in are highlighted.

Net Fishing and Use of Poisons

Net fishing was widely practiced by women in the past. Since the introduction of large gillnets, men have practically taken over the netting of fish.

Traditional netting methods used by women varied slightly and each variant had its own name in Fijian, for example dukuduki, tataga, veitauri, silisili and qoli vakavanua. Because most of these activities were confined to inshore areas, scoop nets *(lawa cua)* and hand nets *(taraki)* were most commonly used. Historically, net fishing was conducted by groups of women who located fish from a distance and then crept close to the schools without disturbing them. To do this, women had to be familiar with the nature of the sand and reef flats, rocks, passages, coral outcrops, mangrove areas, depressions and other features. They closed off escape routes and herded the fish into areas where they could be easily caught.

Women generally used *duva* or fish poison when netting. At low tide the root of a poisonous plant was pounded on the coral flats and then squeezed into tidal pools. Fish trying to escape the poison would get trapped in waiting nets. Because this method kills fish indiscriminately, the use of poison has been banned nationally.

Men's traditional net fishing activities were carried out only on special occasions. Net fishing involved many rituals and employed traditional nets. The men, like the women, had various methods each with its own name e.g. *vakaravi, vola* and *lawa tabu.* Only the *kai wai* or master fishers employed these methods. Most netting activities were similar in nature to the current use of gill nets in that they targeted fish that travel in schools.

Fishing with Barriers and Fences

Women traditionally used barriers or fences for fishing. Stone fish traps *(moka)* or fish fences made from reeds, tree branches and mangrove stalks *(ba ni ika)* were constructed on inshore sand flats. For the *ba ni ika,* an area of about 30-60 square meters was surrounded and secured with vines, with small openings facing the shore. At high tide the enclosures were submerged and as the tide receded, fish were trapped inside. Women then used hand spears, nets and bare hands to catch the trapped fish.

The use of *moka* (stone weirs) was widely practiced in Fiji but has fallen into disuse. A *moka* consisted of a semicircular wall of rocks that closed off a portion of an inshore reef. The openings into the moka were on the shore side to prevent fish from escaping seaward at the turn of the tide. *Tusaliali,* another form of barrier, was erected in an opening in a reef. The fence was circular with its opening to the shore. The barrier was usually left for a day or two before women fished with nets or bare hands within the enclosure at low tide.

In a variation of barrier fishing called "enclosure of rivers" *(latilati ni dreke ni wai)* the construction was similar to *tusalisali,* but in this instance the fish fence was built in an estuary *(gusu ni wai)* or mangrove area *(vanua veidogo).* Fences were usually oval or semi-circular in shape. The barriers were constructed to trap fish that came in with the tide to feed in the estuary or those that were swept downstream toward the sea by floodwaters.

In Totoya, women continue to maximise the use of their immediate coastal area by building fish traps and fences around the reef passages. There is a special fish trap called the *tumata* that involves rocks built up on both sides of a small passageway through the inshore reef. At the time of the incoming or outgoing tide, women wait with their nets to catch fish at these openings. Sometimes, people run through the shallows to chase fish into the waiting nets.

The *yavi rau,* or leaf sweep, was a traditional communal fish drive employing leaves attached to a long vine as a barrier. *Yavi rau* involved the participation of men, women and children (see Veitayaki & Vesi, this volume).

Cina (Nocturnal Torch Fishing)

Both men and women undertake night fishing using torches or lights *(cina).* In most cases fishing targets include a range of species including finfish, eels, octopus, sea cucumbers, shellfish and crabs. In the past, thick bundles of reeds or coconut fronds *(bawara)* served as a torch *(cina bawara).* These days, night fishers employ benzene pressure lamps and waterproof flashlights.

Vakacakau (Gleaning)

The major fishing activity of women is gleaning or *vakacakau.* The term gleaning suggests activities that are unspecialized and of minimal importance. In fact, *vakacakau* requires a great deal of knowledge about habitats and the behavior of species. It also requires skills in identifying shelters and picking up minute trails or signs of the presence of a given species. Women's cautious approach when stalking their prey, and their dexterity and nimbleness when pouncing on different species are skills mastered over many years of practice. The ways in which different species are caught, held, tied and stored is information that fisherwomen have perfected over generations. The ability of women to identify and differentiate edible from non-edible species out of an enormously diverse suite of species speaks of women's vast and often unrecognized fishing knowledge.

Reef gleaning, which is conducted in both shallow water and on the dry reef flats during low tide, refers to all fishing activities which require the prying, digging, cutting, picking up, collecting or breaking off of a wide range of marine food species on the inter-tidal flats and reefs.

This woman has gleaned in mangroves and on mud flats to find crabs and shellfish to sell in the market.

Gleaning is also conducted in mangrove forests. In the murky waters of the mangrove swamp, *qari* or green mangrove crabs burrow into the mud and are identified only by their eyes, which protrude above the surface. In the mangrove vegetation, women hunt them by following the trails that crabs leave on the dry muddy surface. Usually they are found in holes or depressions.

Vivili is a term commonly used to describe gleaning for shellfish. In the mangrove swamps, mangrove oysters or *dioniveitiri (Crassostrea mordax)*, which are cemented onto mangrove trunks and rocks, are dislodged by striking them with a rock. *Yaga (Lambis lambis)* are usually collected at low tide from dead coral beds, dry reef flats and seagrass beds *(veivutia)*. Their presence is detected from a two-line trail on the sand. The shellfish are quite easy to find, as they usually lie partially covered on the nearly dry surface.

Women dive for certain types of shells such as the clam *katavatu*, which is found at the edge of reef slopes adhering to rocks or coral. Women dive using diving goggles and the *katavatu* are pried from rocks using crowbars or steel rods. Tapestry turban shells or *matakarawa (Turbo pethiolatus)* are usually found in outer lagoons or reef areas, at depths of 8 to 15m. *Matakarawa* are pried with steel rods from between rocks where they are firmly lodged.

In Verata, gleaning activities also include the collection of species such as seaweed, sea cucumbers, octopus and jellyfish. These seasonal collecting activities, generally termed *yabe,* are only conducted at low tide when women (and sometimes children) can walk across the reef flats. Sea cucumbers may be collected on dry reef flats or dived for in waist deep water. An octopus is pursued by following the trail that it leaves when it propels itself over the sand. After an octopus is found it is poked *(cula)* on the head with a sharp steel rod *(moto mata dua)* or stick. When disturbed, the octopus stretches and grips whatever it contacts. The fisher immediately grabs the tentacles and the octopus is pulled out of its hole. Unlike the women in Ucunivanua who mostly use crowbars to break up octopus lairs, women in Totoya use sticks.

Line Fishing

Line fishing using nylon hand lines is a common fishing activity of women. Lines are thrown from the shoreline at high tide, from punts and boats anchored over reefs or lagoon areas, over the edge of reefs, or from where women stand waist-deep or knee-deep in water. Women may even fish while swimming in deeper waters of the lagoon. Some use snorkelling equipment but others do not. When a fish is sighted the woman throws her line at it. This method of line fishing is called *basikeli* or bicycle, because a woman swimming with legs suspended in water looks like she is riding a bicycle.

The seaweed Caulerpa (nama) is a traditional food collected by women and sold in Fiji's urban markets.

Other popular variations of line fishing also have particular names, e.g:

- *siwa tutu:* when women fish over reef flats at high tide while standing on punts. When in boats, women sit to avoid being noticed by the fish.
- *siwa qalo:* when women fish standing in waist deep water, constantly changing locations or walking along the reef edge.
- *siwa matu:* when women stand in waist deep water near mangroves or other fish feeding areas and throw out very light fishing lines (2-3 meters long). Because the lines are without lead weights, the fish snatch the bait at the surface of the water and are quickly pulled in.

CHANGES IN FISHING METHODS AND TARGET SPECIES

In Verata, the participation of men and women in fishing has not changed much over time, perhaps because of the continuing abundance and diversity of marine products in the area. The success of local fisheries management initiatives in reef and lagoon areas in recent years has ensured a steady supply of marine species. This allows women to steadily meet family and economic demands, and may explain why women still fish within their traditional fishing areas.

Many of the fishing techniques discussed above are still in use but in some cases women have lost the knowledge associated with such fisheries. Women's fishing activities have thus changed in the sense that some traditional activities (netting, crabbing, night fishing, chasing fish on dry reef flats) are now not as common as they were in the past.

Although in some parts of Fiji, modern equipment has replaced traditional fishing tools, in Verata women's current fishing practices are in most cases modifications of traditional fishing techniques with minimal technological change. It is a common sight to see women out on the dry reef flats on fine days, walking, collecting and searching for edible shellfish, seaweed and other marine species. Armed with knives, sticks and at times crow bars, women cluster in groups in certain areas of the reef, prying at rocks, netting in the small tidal channels, collecting seaweed or sea cucumbers or digging for sea shells. Other women can be seen going off in groups in small powered boats to more distant reefs and other fishing areas. These women are often fishing for species to sell at the municipal market.

From a technological perspective, therefore, most of women's subsistence fishing activities have remained the same, e.g digging for shellfish with their hands and feet, using sticks and knives to dig up and cut sea cucumbers and other species, and line fishing along inshore reefs. Women's fishing activities have, as a rule, not benefited from improvements in fishing technology because of the nature of their fisheries and the species that they target.

What has changed most is the focus of fishing. In the past, people fished for food and gave away, sold or traded the surplus. Today the catch is basically for sale, with the surplus being kept for family consumption. There is a marked difference between women who fish commercially or for subsistence, in terms of the allocation of fishing time. For example in Totoya, which has no access to markets, women's fishing activities are still very flexible and selective in nature. The species targeted depends principally on the season and the weather, and not on market demand. For instance when the south-east trade winds blows, the women come out in force to search for octopus.

Where markets are within reach, for example in Ucunivanua, many of the women's fishing techniques may have remained the same but their fishing targets, the time and effort spent in fishing and how they distribute the catch have changed considerably. Fishing activities have always varied to some extent from place to place. Some women specialise in those species that are most common in their fishing areas. Others focus on the fisheries they were exposed to as children. But, more recently, fishing activities have also been guided by proximity to markets. Women living near urban areas have tended to shift their efforts to the species that fetch good prices. In some cases, species that are not consumed locally are harvested for selling at the markets. Fishing patterns and target species are determined more and more by the market price or by consumer demand.

Increasing needs for cash at the village level have pushed some people to develop fishing enterprises based on family units. It is not uncom-

SOCIAL IMPORTANCE OF WOMEN'S FISHING

There is a common, modern misconception that women fishers are under a lot of pressure and stress to get fish for the family. My mother, who is an avid fisher, was taken aback when I once asked her if she did not think that fishing was a lot of work with very little reward. She and her friends would spend hours on the banks of the Rewa River in Fiji where I grew up, line fishing for *tilapia* or grass carp. Fishing was not only getting food for the family, but was also a test of skills and an opportunity to be together with other women and catch up on village gossip. To them, fishing was a release from the confines of the home.

In Totoya, after the first few catches, women settle down in groups and cook their catch over open fires *(tatavu)*. They enjoy lunch before they resume fishing for food to take home.

The social dimensions of women's fishing activities are usually not considered when recording the participation of women in fishing.

mon to see men, women and children participating in certain fishing activities. Also, women are increasingly fishing beyond the immediate shore areas and are participating in what were previously defined as men's fishing activities. There is more reliance on modern fishing technologies and people are moving farther out to fish deeper waters for certain desirable species.

FISHERWOMEN'S NETWORKS

Despite the changes noted above, women's preference for fishing in groups is still strong. In Verata, women fish in groups when hiring boats, fishing for a pre-placed order or when gleaning. It is usual, for example, to see women out on the reef flats in small groups rather than as individuals. Through these groupings, women have built networks, which they exploit to assist them in their fishing activities. Marketing is also organized in groups, with each group in turn taking other women's catches to market on the weekend. When interviewed, some women engaged in fishing and selling at the local markets said they could not have done so without support from other women in the village. The networks that women have organized for going fishing or marketing relieve some of the burden of their participation in commercial fishing.

POST-HARVEST ACTIVITIES

Post harvest processing of marine species is almost exclusively the responsibility of women. Post harvest activities usually involve traditional methods such as smoking and drying. Octopus are trampled on in sand or beaten with pawpaw stalks to soften the flesh, before being hung up to drain out the liquid before smoking. Various smoked and dried species are sold at municipal markets. In other cases, women use their traditional skills to keep marine products fresh for the market. Shellfish and clams may be kept in small tidal pools along the beachfront adjacent to the village until the day before they are taken to markets. Seaweed is wrapped in banana leaves and kept in tightly closed bags. In Totoya women also temporarily store their shells in the tide pools that appear at low tide. For example, clams and other shellfish are kept in tidal pools that are lined by stones. Other women know who these catches belong to and no one touches them.

DISTRIBUTION AND SALE OF FISHERY PRODUCTS

The distribution and sale of marine products is almost exclusively the task of women. In most cases, women take charge of the catches of all the members of the household. There are also urban traders who place orders and buy in bulk from the village women as soon as they get to the markets.

In Verata, the finfish that men catch are sold through municipal markets. Women supply the shellfish, crustaceans, smaller finfish and other miscellaneous products. Because women regularly go to market there is increased demand for the capture of species that were previously not harvested or underutilized.

Changes in the participation of women in fishing are most obvious at the market. A day spent at any of the municipal markets in Fiji reveals that compared to the past, many more marine products are now sold. Women have also, through the years, perfected harvest and post- harvest techniques to meet evolving consumer demands. For example in Fiji, the availability of ready-to-eat seafood has steadily increased. Women are selling certain species already cooked or prepared raw. One woman who was selling shelled mud lobsters at the Suva market said that, even though it was harder work to sell cooked food, there was a ready market among workers who do not have time to prepare seafood while on the job. This demand for the product was enough to convince her to keep on cooking her lobsters. Both men and women trapped these mud lobsters but only women took the product to market.

Smoked fish being sold at the market.

In the Suva market, sellers are from various provinces and many rural coastal communities. One female fish seller interviewed explained that her husband travelled weekly to the other main island, Vanua Levu, to buy seafood. The clams and other sea food were shipped to the woman in ice boxes, and she sold the products from

Thursday to Saturday of each week. Some vendors travel by boat from small islands and then catch buses or carriers (open trucks) to Suva. People who had come in to the Suva market from Moturiki and other outlying islands in the Lomaiviti group said they had arrived the day before and slept either with relatives or outside of the market, waiting to sell their products (mostly crabs and land crabs). Women from Verata, Namara and Kaba who were at the Suva market stated that they preferred selling in Suva over Nausori because they were guaranteed better sales in Suva. These women start coming in to the market as early as midnight on Fridays so they can secure a good selling spot. They spend the whole night in the cold outside the Suva market and return home only after selling their products, which is usually late the next afternoon.

Many techniques and routes are used to distribute marine products around Fiji. Rural women have made inroads into urban markets. Women from the western islands and villages of Fiji are selling marine products in the east and vice versa. This two-way trade involves not only marine products but also land products and food crops. During the orange season, oranges from the east flood the western markets. During the mango season, the western mangoes flood the eastern markets.

A steady stream of seafood such as sea grapes, fish and clams which are sold at the Suva market are from the Yasawa Group that lies to the west of Lautoka city. Women from the Yasawas also sell weekly at the Lautoka market. In some cases, Yasawa women living in Lautoka act as brokers who order and purchase seafood from the islands. In other cases, Yasawa women fish, preserve their catch, transport it to the mainland and board with relatives in Lautoka while they sell their products. Much of the abundant seafood in the Lautoka market, including seaweeds, clams, crabs and smoked fish, are from the Yasawas. Fish is smoked because of poor transportation services and problems with freezers. Most of the women interviewed had to leave Yasawa on Thursday and sell products on Friday and Saturday before returning home.

In Labasa (Map 1) on Vanua Levu, fish is sold fresh almost daily by small-scale boat operators from nearby villages and by commercial boat owners, but women sell seafood only on Saturdays. Because of transportation problems, the women on the north-eastern coast of Vanua Levu can come to the town market only once a week. Fish sold by them are mostly smoked; they also sell crabs and shellfish. Women from Udu, who are the usual sellers, come into town on the morning bus and return home on the afternoon bus on the same day. There is also a lot of fish sold to middlemen in Labasa.

CHALLENGES FACING FEMALE FISHWORKERS

The participation of women in fishing and their substantial contributions to family livelihoods are not fully acknowledged. Rather, their efforts tend to be taken for granted. Even commercial activities are viewed as being just a part of their everyday work, and are not recognized as independent economic activities. Women at work in the sea are said to be "just gleaning". Their understanding of species, habitats, seasons and other factors which affect fishing is undervalued.

Although women are increasingly active in marketing resources there are no support institutions to assist them. Because no adequate credit facilities are available, women take out unsecured, small amounts of credit from shops. Thus there is no opportunity for long-term planning. Consequently, women are locked in a cycle where they fish to pay off credit and have no real profit at the end of the day. In addition, the need for money pushes women to sell off most of their catch. At the end of the market day, they buy cheap meat and tinned food for the family. Thus, there is a deterioration of nutritional status in areas where people rely significantly on marine resources for their economic livelihood.

Transportation to markets often runs at irregular hours rather than to schedule. The vehicles are often very unsafe, as trucks are usually overloaded and lack safety devices such as proper seats with seatbelts. Because they live in rural coastal villages, travel to the market could take three to five hours on these trucks. In order to secure good selling locations, women leave home early in the morning and as a result lose proper sleep and meals. Because of the continual activity of fishing and selling, women can over time develop nutritional and health problems. As women spend more time on fishing and marketing, the nutrition and health of their families also suffers. Older women or sisters may be recruited to help look after the families of vendors. Thus the burden of increased fishing and selling again falls on women.

Women who sell cooked meats would benefit from training and improved facilities.

IMPROVING THE FUTURE FOR FISHERWOMEN

Women have, on one hand, attained empowerment and independence through participating in fisheries, but on the other hand, they face many constraints and difficulties. There are several areas where governments and women's organisations can assist the fisherwomen of Fiji.

Institutions and networks that are used by women as coping strategies at the village level should be recognized and used by development practitioners working with fisherwomen. Rural networks could provide enhanced avenues for trade. For example, instead of women working through middlemen in distant markets, they could exchange or buy marine products from other women's groups.

Such networks could also facilitate the transfer of information and technologies related to improved fishing, processing and preservation. The networks can be used to facilitate awareness and education work that specifically targets women. Improvement of their knowledge base could assist women in addressing problems related to the use of destructive fishing methods, over-harvesting, and long-term impacts of resource depletion.

Transportation to markets and market conditions require serious attention from decision making authorities and community groups including the Fijian Administration, women‘s NGOs, Women‘s Interest Officers, Municipal authorities and business organisations. The introduction of technologies to ease work and improve the health of women engaged in harvesting and processing should also be seriously considered. The traditional knowledge, skills and understanding that women have of their fisheries, should be documented and preserved. Such knowledge will be useful for the sustainable use of resources and could enhance coastal management initiatives.

Proper evaluation of the subsistence fishery is essential. In particular, the participation of women should be accurately assessed to provide data for consideration by those who lead fisheries development efforts. Availability of accurate data on their commercial activities could also support the bid by women fishers to be considered for loans and small-scale financial assistance.

Proper training and assistance should be provided to women engaged in the expanding trade in cooked seafood so that the foods meet national health standards. There is also a need to educate women about the value of the fresh foods they sell and the relatively inferior nutritional quality of many of the foods that women purchase with their fishing incomes.

The Department of Women and Culture and women‘s NGOs in Fiji could work towards setting up a national fisherwomen‘s organisation that would include fish sellers and look after the welfare of female fishworkers. Another area requiring attention from such an organisation is how to deal with surpluses in sea food so that these can be preserved for sale.

GENDER EQUITY AND GLOBALIZATION

This case study shows how women's participation in fishing has expanded to include post-harvest, distribution and marketing activities. Because these activities are poorly documented and unrecognized, women have difficulty organizing themselves officially as fishworkers. Their fishing activities are considered a part of their household chores rather than as serious economic ventures. As a consequence, women are still expected to attend to all their domestic chores even when fishing and marketing consumes large portions of their work week. In traditionally organised, rural Fijian societies, the roles of men and women are defined, and women, despite their many other activities outside their homes, must meet traditional and household obligations.

Globalization has forced fishing communities to develop strategies for securing their livelihoods. Women rely on their village groupings and networks to survive in the competitive market economy. Women still lack negotiating power because they are excluded from decision making at the village level. They also lack control over assets that might enable them to set up proper economic ventures. Yet, in spite of all these challenges, women function as "invisible participants" in the modern market economy in order to meet obligations to provide for household food security. For this situation to change in the context of a strongly patriarchal culture, governments will have to set in place an enabling legal environment that recognizes, protects and strengthens the rights of women in Fiji.

REFERENCES

Lal, P. N. and C. Slatter 1982. The Integration of Women in Fisheries Development in Fiji. Report of an ESCAP/FAO initiated Project on Improving the Socio-Economic Condition of Women in Fisherfolk Communities. Suva: Fisheries Division, Ministry of Agriculture and Fisheries & Centre for Applied Studies in Development.

Matthews, E. 1993. Women and fishing in traditional Pacific Island cultures. Workshop on People, Society and Pacific Islands Fisheries Development and Management: Selected Papers in Inshore Fisheries Research. Project Technical Document 5:20-33. Noumea: SPC.

Ram-Bidesi, V. 1993. Women in commercial fisheries in the South Pacific: A focus on the Fiji situation. In R. South (ed) The Ray Parkinson Memorial Lectures: Marine Resources and Development. Suva: PIMRIS, University of the South Pacific. P.105-119.

Rawlinson, N. J. F., D. A. Milton, S. J. M. Blaber, A. Sesewa, S. P. Sharma 1996. A Survey of the Subsistence and Artisanal Fisheries in Rural Areas of Viti Levu, Fiji. Suva: Fisheries Division.

Schoeffel, P. 1984. The fisheries of the South Pacific: Strategies to assist the participation of women in fish marketing, small agriculture and fish processing. Draft Report, Pacific Multi-country Women in Fisheries Project. Ocean and Shoreline Management. 15(1): 309-324.

Slatter, C. 1984. Women's Role in South Pacific Agriculture. Traditional, Transactional, Modern Roles of Women with Special Focus on PNG, Fiji, Western Samoa, and Tonga. Traditional Knowledge Utilisation Project. Honolulu: Institute of Culture and Communication, East-West Centre.

Slatter, C. 1995. Changes to women's roles in fisheries development in Fiji. In E. Matthews (ed) Fishing for Answers. Women and Fisheries in the Pacific Islands. Suva: Women and Fisheries Network. P.71-90. Thaman, R. 1982. Deterioration of the traditional food system, increasing malnutrition and food dependency in the Pacific Islands. Journal of Food and Nutrition 39 (3):109-121.

Talagi, S. 1989. Report on the Role of Women in Small-Scale Fisheries in the South Pacific. Case Studies in Cook Islands, Papua New Guinea, Solomon Islands, Tonga, Vanuatu, Western Samoa. Food Production and Rural Development Division, Commonwealth Secretariat.

Veitayaki, J. 2000. Fisheries Development in Fiji. The Quest for Sustainability. Suva: Institute of Pacific Studies and Ocean Resources Management Program, University of the South Pacific.

Vunisea A. 1994. Traditional marine tenure at the village level: a case study of Ucunivanua, Fiji. In G. R. South, D. Goulet, S. Tuqiri & M. Church (eds) Traditional Marine Tenure and Sustainable Management of Marine Resources in Asia and the Pacific. Suva: International Ocean Institute-South Pacific, Marine Studies Programme, USP. P.200-207.

Vunisea, A. 1995. Subsistence fishing, women and modernisation in Fiji. In E. Matthews (ed) Fishing for Answers. Women and Fisheries in the Pacific Islands. Suva: Women Fisheries Network. P.101-108.

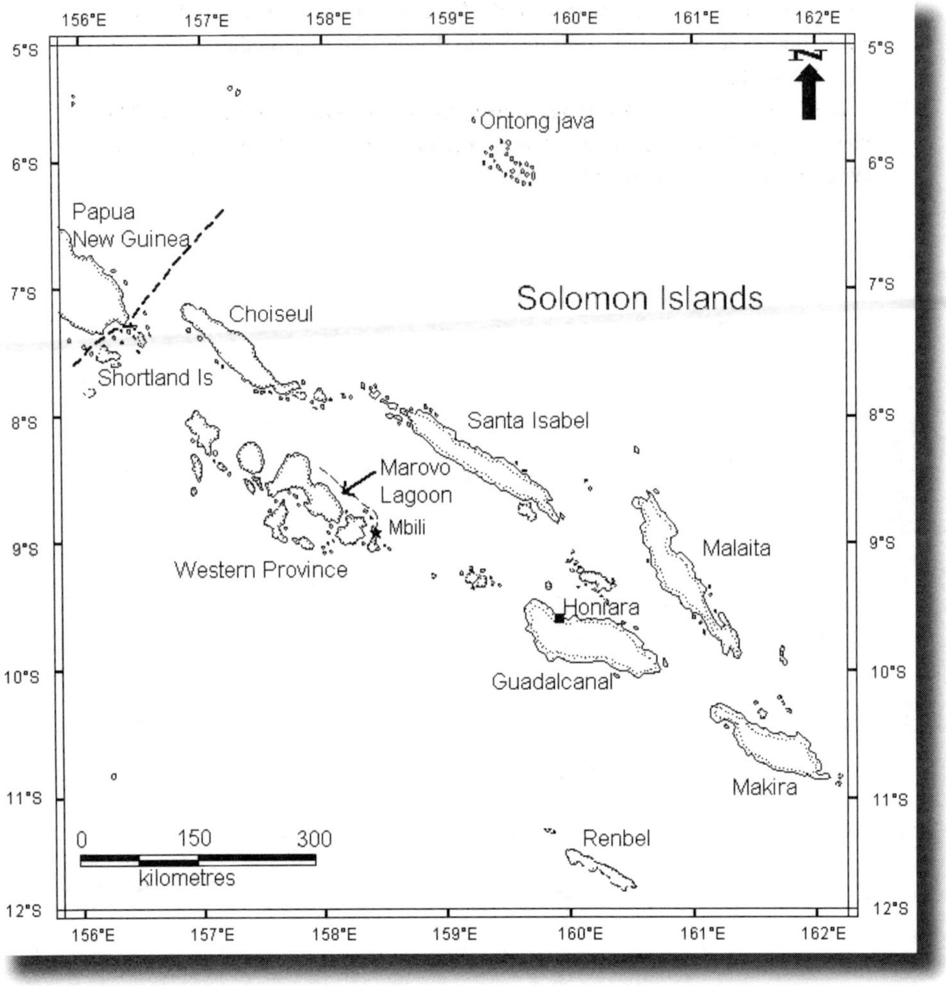

Map 1. Solomon Islands, showing location of Marovo Lagoon in Western Province.

8 Feast or Famine? Fishing for a Living in Rural Solomon Islands

Alan Sepe Agassi

"Marovo lagoon is one of the biggest lagoons in the Pacific region and is considered to have plenty of marine resources, including high quantity and quality bait fish stock...."

— Hansel Mamupio, Manager of Noro fishing base for Solomon Taiyo Company

"There were so many bait fish types in the reefs, lagoons and also out in the ocean deep. In the past some of these bait fish came ashore; people collected them for food. Today, in particular after the arrival of bait fishery activity in Mbili, there is no more bait fish coming ashore."

— Johnson Poghoso, Chief of Mbili village, Marovo Lagoon

PERCEPTIONS OF CHANGE IN INSHORE FISHERIES OF THE SOLOMONS

Somewhere between the perception of the chief of Mbili village and that of the manager for the bait fishing company Solomon Taiyo there lies the truth of the state of fisheries in Mbili, a small village on the edge of Marovo Lagoon in Western Province, Solomon Islands.

This case study will explore the history of Mbili's fisheries, and the factors that have shaped the relationship between Mbili people and the Marovo Lagoon fishing grounds. One of these factors is the bait fishery carried out since the late 1970s on the traditional fishing grounds of Mbili. The positive and negative impacts of this industrial fishery over time, and its relative importance compared to other factors affecting local marine resources, will be documented. Other factors relevant to marine resources and their management include traditional marine tenure arrangements, chiefly leadership, changing technologies and gender roles in the fishery, the influence of the church, increases in population, the evolving cash economy and the recent history of ethnic conflict in the Solomon Islands. I will document, through the testimony of Mbili people, how the marine resource base has been affected over time, and present ideas of how the future of fishers in Marovo Lagoon can be secured.

CASE STUDY METHODS

Data for this case study were gathered through a combination of literature search, observation, focus group discussions and key informant interviews conducted in Mbili, Noro and Honiara, Solomon Islands, over the period August 2003 – January 2004.

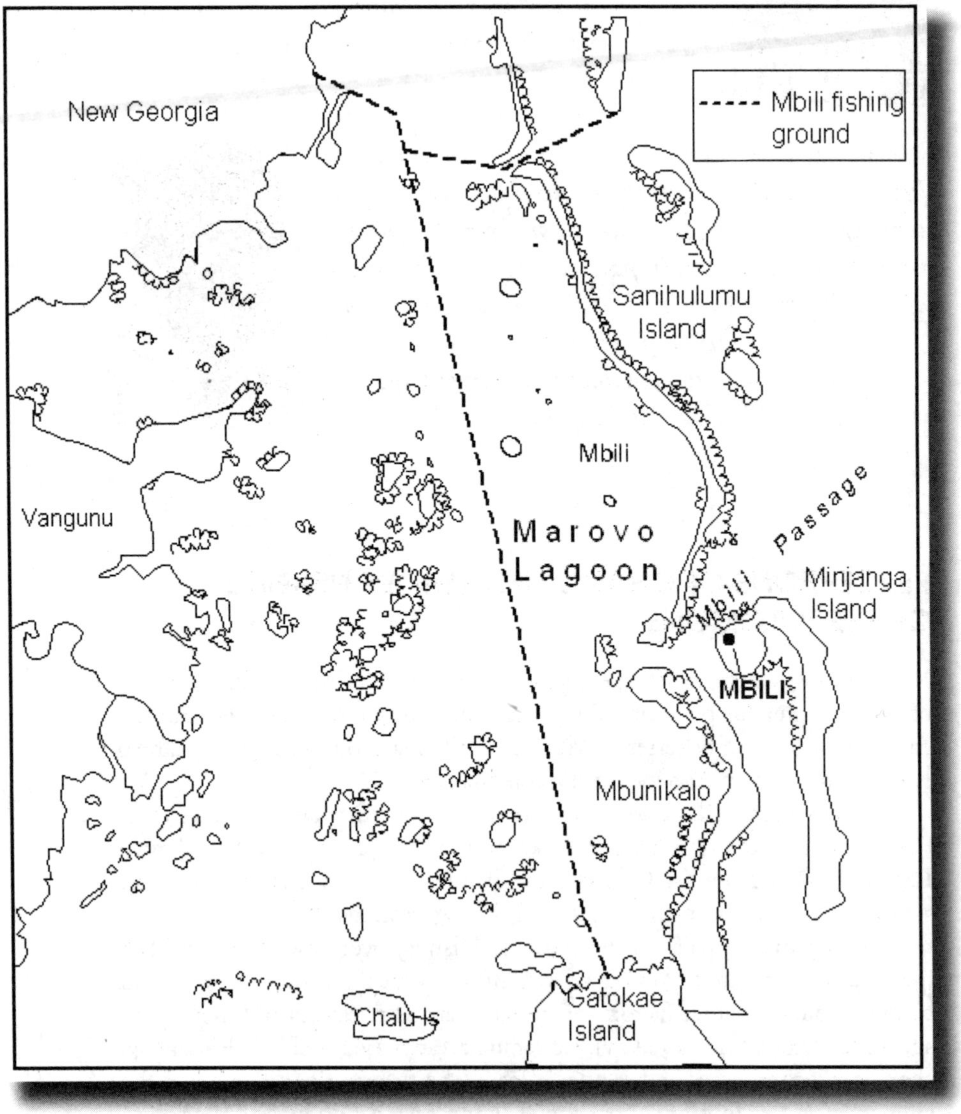

Map 2. Mbili traditional fishing ground

MBILI VILLAGE

Western Province in Solomon Islands is made up of seven major islands and more than 500 smaller islands, which support 62,000 people (Statistics Office, 1999). Prominent geographic features of the province are the large Marovo and Vonavona lagoons. Mbili village lies on the tip of Minjanga Island, overlooking the main southeastern entrance to Marovo Lagoon. The lagoon runs along the north-eastern shores of the main islands of Gatokae, Vangunu and New Georgia (Map 1).

Mbili people have not always lived in this place. When the Seventh-day Adventist Church decided to build a mission at the present location of Mbili in 1916, the people moved in from Govara Kiso on the other side of the island. In 1995 there were 31 households and a population of 179 people in Mbili (Statistics Office 1997). Since that time the population has increased and some families have moved out to six settlements on adjacent islands. There are neither roads nor motor vehicles in Mbili, nor is there electricity or any septic system. Beaches and mangrove areas serve as toilets. Rainwater collected from the roof of each house is stored for domestic use. The church continues to play a prominent part in the daily life of the village.

The main source of income in Mbili is wood carving, which occupies most of the male population (Kupiainen 2000). Before conversion to Christianity, carvers made fighting clubs and axes; these days they carve souvenirs for tourists. Women earn some income from weaving mats, hats and baskets. As is true for rural areas throughout the Solomon Islands, most families subsist by fishing and gardening (Otter 2002). Fish and garden produce that are in excess of a family's needs may be shared with the extended family, sold locally or sent on ice to the market in the capital city, Honiara.

The land territory attached to the village of Mbili covers not just Minjanga but also at least nine other islands, several of which are larger in area than Minjanga (Map 2). The land is divided among various clans descended from the original settlers. In addition to access to plots of land for agriculture, all villagers have free access to a large marine territory that is bounded by Tongoro Passage in the west, the shipping channel through Marovo Lagoon to the south, and the mouth of the Kumabusi River to the east. The seaward edge of the marine territory is not well defined.

Local Governance

In the village there are three chiefs. One looks after land and sea properties; the other two are responsible for social conditions and community organizing. All three are recognized owners of the marine territory, with ownership rights descending through both male and female lines. When elderly or ill, a chief nominates his successor in consultation with the other chiefs. The successor must be from the chief's immediate family (i.e. brother, son or - if these are absent - the male child of a sister). A local

THE LEGEND OF IGA

Iga, the chief of Toba and his family were the first to come and live on the islands around Gatokae. Iga made these islands inhabitable, as they were once infested with demons, bad luck and evil spirits so that no man could live on them. One day, headhunters from Russell Islands and Isabel found Raptali, the wife of Iga, sitting at the bottom of a casuarina tree at the seashore, wearing her custom jewellery called *Hokata* and *Tinete*. They tried to steal the jewellery but she broke them, so they killed her. Further along the coast they found Iga at Tuva Tusu with his child Meha. The warriors recognized that he was the chief and one of them said "so he has no blood" [he cannot be killed]. This warrior went down and took the baby Meha from Iga's arms and cut Iga's head off. The raiders took captive three of Iga's children: Koleki, Getu and Meha.

Raeboe Logara, Mbili chief.

When the headhunters rested on the island of Kicha a retaliating group from Marovo caught them. They quickly ran away and left Meha behind on the island, where he was killed by the Marovo warriors. The two remaining children were taken to Isabel until they became mature. Getu, because he was male, decided to return to his original place of Gatokae. So he went back and lived on the islands.

Upon his return, the people of Marovo gave Getu a wife. Because Iga shed his blood to protect the people of Matagele, they also gave Getu the right to own the islands and the adjacent reefs and lagoon in compensation for the death of his father. "And that is why" says Raeboe Logara, chief of Mbili "we now own these islands and marine areas."

legend speaks of how, in ancient times, the chiefs of the Mategele tribe granted ownership of these fishing grounds to the son of a warrior named Iga. Thus, all of today's chiefs are descended from Iga.

Sometimes the chiefly right of reef ownership is contested. In 1986 descendants of a stepdaughter of Getu (Iga's son) claimed ownership of the Mbili reefs. The chief at that time was descended from Getu through Getu's Marovo wedded wife. The contenders were descended from Getu's wife from Isabel. Because it was claimed that the Isabel woman's child was actually fathered by her other husband in Isabel, and not by Getu, the challenge was defeated.

TRADITIONAL ARTISANAL FISHERIES IN MBILI

Traditionally, the people of Mbili fished in the lagoon directly in front of their homes. In those days "living style was very good; there were not so many people" and there was no selling of fish. Rather, "village people shared fish they caught to all houses" (Euzene Hungo, Mbili elder). The efforts of just a few fishermen were enough to feed the entire village.

Traditional Fishing for Reef Fish and Skipjack Tuna

Before nylon fishing lines, spear guns and dynamite, reef fishing in Mbili was carried out using a variety of traditional gears and methods (Table 1). Women typically fished without specialized equipment. Fishing with spears was popular among Mbili men and spearing fish was easy because fish were plentiful. Various hook and line techniques as well as traps were also employed. Over time, hand spears were replaced by the earliest form of diving gun.

Before the industrial tuna fishery began, villagers fished for skipjack tuna (Table 1). Men did the fishing but women helped to paddle the canoe (Niba Logara). "The skipjack tuna… was shared out to every household in the village. They would count every person in every house in the village present at that time and cut it up to share equally" (Johnson Poghoso). Today, this traditional tuna fishery has been abandoned.

INDICATORS OF CHANGE

The past richness of the fishing ground is reflected in comments by elder Niba Logara, who recounts that in the past, before a woman would go fishing she would already make a fire. Then she would walk down to the beach and fish and within a short time she would have caught enough for the family meal. Today, Kathleen Johnson faces a different reality. She must travel farther and fish for longer, and make her cooking fire only after returning from the sea.

Niba Logara, Mbili elder.

Kathleen Johnson tends her cooking fire

Table 1. Traditional techniques for catching reef fishes and tuna.

Local name	Description
Kerukeru	Chasing fish into holes and catching them by hand, performed by women (men's focus group)
Buna Buna	Leaves of a poisonous plant were pounded with sand and placed into holes in the reef to kill the fish and make them easier for women to collect (John Ago, Mbili fisherman).
Dumi	A hand held spear used by men to catch fish while diving, or thrown from a canoe. For night spearing, a man in a canoe used a burning torch to sight fish on the bottom (Raeboe Logara).
Hipuru or *Hipuhipuru*	Men threw a baited fishing line into shallow water to catch larger fish. Sometimes the line was attached to a pole (Ellen Soga, Mbili elder). Fishing lines were made from a vine called *pusi* (Johnson Poghoso, Mbili chief). Hooks were carved from local materials such as turtle shell. Different sizes and shapes of hooks were carved to catch different fish and each hook had its own name.
Ururu	Line fishing at night (Niba Logara).
Lobaloba.	Trolling on the surface on a moonlit night, using a baited hook and white lure. Night trolling is still practised today by both men and women.
Dukolopete or *Veko Vekoi*	Drop line fishing in deep water using stone weights attached to a fishing line (Mens focus group and Niba Logara). Modern day fishing when a lure is pulled through deep water is known as *kurakura*.
Kura	Trapping method in which a fisherman in a boat lowered a woven basket over the side. The basket was made of vines and contained bait and a sinker (*doko*). When a fish entered the basket to eat the bait, the fisherman pulled the mouth of the basket shut using a rope called *iloro* (Raeboe Logara).
Loklani	Earliest form of diving gun made of two sticks connected with a piece of rubber. One stick bore a sharp wire on the tip for spearing fish. The rubber was used to propel the spear and to pull it back when the fish had been shot (Raeboe Logara).
Valusa	Fishing for tuna using a string attached to a bamboo pole and hand made hooks called *ghaili* (Meliteko Andrew).

Bait Fishing

Ellen Soga, Mbili elder, with her mat weaving materials.

Mbili people caught and ate a variety of small pelagic fishes known collectively as bait fish. "In the past the people collected them in the lagoon at the right times. They came up to the surface of the sea in large schools and people would collect them by scooping them up using local baskets… The last time I can remember seeing this was in 1975 in Muku" (Johnson Poghoso). Bait fisheries also had their own special equipment and techniques (Table 2).

Kevo, a small fish used by women to bait hand lines, as well as fishes locally called *medarae* were often targeted using hook and line. When *medarae* were fished with the aid of a traditional fish aggregating device (see *rapaka*, Table 2), no women were allowed to walk on the platform or catch the fish; it was taboo (Johnson Poghoso).

Table 2. Traditional methods and gears for catching bait fish.

Local name	Description
Saputu rita	Using a hand held fishing line to jig the small bait fish called *kevo*, an activity still carried out by women today. In the past they used a fishing pole, fibre line and hooks that were called *voleke.* These bait fish were caught in the early morning and late afternoon.
Rapaka	Hooks made from turtle shell and black lip sea shell, used to catch the bait fish known locally as *medarae* (Johnson Poghoso). *Medarae* were sometimes caught with the aid of a platform covered with coconut fronds. The platform gave shade that attracted the fish, which were then caught by men standing on the platform using pole and line with *rapaka* hooks.
Voleke	Hooks made of the same materials as *rapaka*, but used to fish for the bait fish known as *kevo* (Johnson Poghoso).

Communal Fishing Practices

In the past, Mbili men and women worked together in communal fishing activities. Each technique had a special name and would be carried out on particular fishing grounds (Table 3).

Table 3. Communal fishing practices traditionally employed in Mbili

Local name	Description
Ara	A stone fence was built on the reef. At low tide, fish were located by a master fisherman. Local vines called *arara*, with their leaves still attached, were tied end to end to form a long rope. Men and women held this rope and swam behind the fish to herd them toward the fence to be killed. Many people had to work together to get a good catch (Johnson Poghoso).
Balava	Coconut leaves were used to drive any available fish inshore to a killing area (Johnson Poghoso).
Hado epata	Fishing was performed on the reef flat early in the morning (around 3 a.m.). As fish were herded into a killing area using vines, tree branches were placed across their escape routes. This was carried out only at Chapa and is no longer practised (Johnson Poghoso and Raeboe Logara).
Kuarao	Men and women swam in from the deep sea holding between them an *arara* vine to guide any fish that were around into a shallow killing area (Johnson Poghoso). This kind of fishing was done in particular areas in Tungupaga and Tusu te Koli, but has not been practised since the early 1980s (Raeboe Logara).

Traditional Ecological Knowledge

Listening to elders in Mbili tell stories of the early days reveals a significant depth of traditional knowledge. The people tell how they would go to particular fishing areas at different phases of the moon when certain fish were known to be abundant. They know the inshore spawning areas of various reef fish. They also say that in the past, the schools of bait fish were so thick that the men could spear the big predators that swam with the bait before these big fish would even see the fisherman.

THE TRANSITION TO SMALL-SCALE COMMERCIAL FISHING

Over the years the population of Mbili has increased and many aspects of society have changed. Elders interviewed in 2003 noted that more and

WOMEN IN TRADITIONAL FISHERIES

The elders' stories make it clear that, apart from helping in communal fisheries and sometimes paddling canoes for male fishers, women hardly fished in the past and children were not allowed to fish at all. Men were dominant in the fishery, venturing to sea in boats to catch large reef and pelagic fish. Women fished individually or in small groups on a casual basis. "Poison fishing and gleaning of fish from their holes in the reef flat was done [by women] for picnic or camping consumption. They did not always do this to take fish back to the village household; rather it was cooked and eaten out on the beaches or islands or even while working in the gardens" (Euzene Hungo).

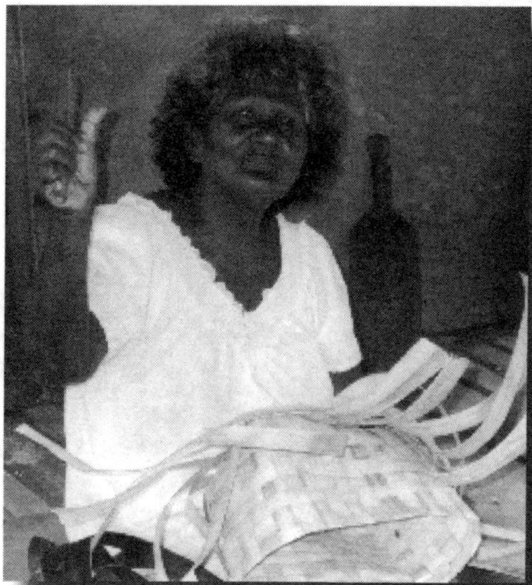

Euzene Hungo, Mbili elder, making a basket for sale

Today the situation is dramatically different. Women not only are fishing, they are also selling their catch and are fishing in deep water as well as close to shore.

more families now live in permanent houses rather than in traditional leaf houses. Children attend school and must pay school fees. The dress and hairstyles of young people have changed under the influence of urban centres. With these changes comes an increased engagement in the global economy and an increased need for cash. Also, say the Mbili elders, societal values have changed. Where once there was great respect for chiefs and elders, in these modern times the traditional hierarchies are breaking down. At the same time, families are becoming more self-contained. Whereas in the past people shared their food and exchanged excess products through barter, now transactions are done with cash, and cash for labour is more common than volunteer service. As the population increases people also have begun to move out of Mbili and into smaller settlements on islands in the lagoon.

One of the earliest cash economic activities was the harvest and sale of topshell (*Trochus*). Travelling traders purchased this highly valued commodity and sold it in the international market. Bêche de mer (sea cucumber) is another valuable species that has been fished and exported from Mbili. The Seventh Day Adventist church has always discouraged bêche de

Solomon Island style leaf house

Fish from many islands comes to the fish market of the capital, Honiara, to be sold.

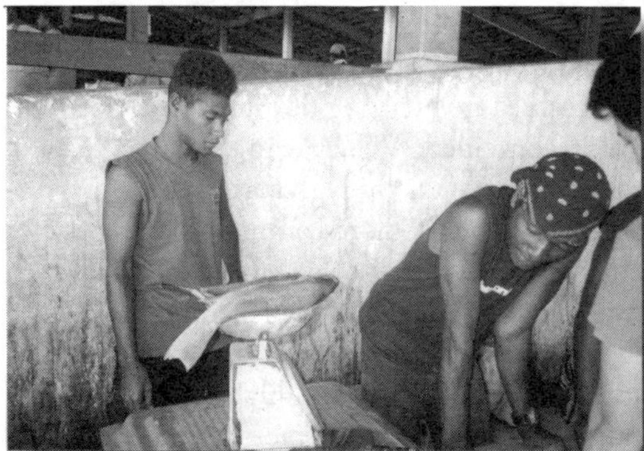

Mike Grymis makes a sale of fresh fish from Isabel Island.

mer fishing. The current fishery is a sign that some people now choose not to follow the church's dictates.

In the late 1960s a wharf for inter-island cargo and passenger boats was built in Mbili. This was later moved to nearby Mbunikalo. By the 1970s, advances in shipping infrastructure made it possible for people to ship fresh fish on ice to the distant market of Honiara. With access to distant markets, catching fish for sale became more important and communal fishing practices were abandoned in favour of individual efforts.

MBILI FISHERIES IN DECLINE

Mbili fisher people agree that it takes longer to find fish now, compared to 10 or 20 years ago. One must go farther out from shore and dive deeper. Whereas in the past men went fishing a few times a week, now they go out every day except Saturday, which is the day they attend church. People also report that many more villagers are fishing. Of particular note is that the number of women and girls fishing has increased greatly. Women now also fish in deep water and at night, and they are fishing commercially, not just for food.

Fishing methods have become more efficient and destructive. Although log canoes are still made and used in Mbili, most traditional technologies have been replaced with modern methods. Nylon lines, metal hooks, gillnets and spear guns have increased the fishing power of local fishermen and women. Fish poisons have always been part of the fishing tradition but whereas the Mbili people would always use leaves, now they have learned from Papua New Guinean fishermen how to use much more toxic plant roots (John Ago, Mbili fisherman). Dynamite fishing has decimated some areas. Live fish traders targeting large reef fish have also moved through Mbili, so that large reef fish are now rarely encountered. Bêche de mer are scarce due to over harvesting and some people fear that their depletion may be affecting other species. Some villagers say that spear gun fishing at night has exhausted many reef fish, potent poisons have killed patches of coral and logging in the watershed has spoiled fish habitat.

Mbili woman with her daughter, beside their fishing canoe.

Mbili men returning from a successful fishing trip. Left to right: Lidon Songa, Irving George, Willy Posala, Garry George and Fox Ata.

Small pelagic fish stocks also appear to be in trouble. There are reports that the bait fish called *tingini* has disappeared from inshore waters and that two others, *inabuku* and *kevo,* are much less common than before (Johnson Poghoso). According to village chief Johnson Poghoso, the *ghalusu*, a relatively large bait fish, and *moturu*, a predatory fish that chases bait fish inshore, have been wiped out by dynamite fishers.

Fishermen report that in general, the bait fish schools are much thinner than in the past and it is harder to catch the big fish, including skipjack and kingfish, that run with the bait. "Because the bait fish are gone from around the inshore stones where they used to gather, there are no longer any big fish to catch in these areas" (John Ago).

When fish resources decline, traditional knowledge associated with them may also be at risk. In Mbili, people used to know where different kinds of fish could be caught at different phases of the lunar cycle. This traditional knowledge is becoming useless and lost because "It seems that seasonal fish now come out at any time and not according to the season known by people in the past" (Fishermen's focus group). The Mbili fishermen have also noted bleaching of some coral reef areas, a sign that rising sea surface temperatures and changing weather patterns related to global climate change may play some role in the changes they have noticed in fish migratory behavior.

The increase in fishing pressure and the perception that both reef and pelagic fish resources are in decline causes concern among elders and young people. Yet, rather than admit that the fish may in fact be depleted, some maintain that their lack of success in fishing is because somehow the fish today are smarter and that is why they are hard to catch. Others surmise that because of coral bleaching and other habitat damage, as well as increased activity of people in shallow waters, fish have moved to "a better place to live and find food", in deeper water.

THE INDUSTRIAL BAIT FISHERY OF MBILI

Bait Fishing Activity

Japanese business interests established a pole and line fishery for skipjack tuna in the Solomon Islands in the late 1960s, in large part due to the availability of small pelagic fishes that their boats require as live bait. "When the Japanese company surveyed the bait fish stock in Solomon Islands in 1971-72 they found that Marovo Lagoon had the biggest bait fish stock of anywhere in the world, so the Solomon Taiyo Limited decided to establish pole and line fishing in the country" (Hansel Mamupio, Soltai Export and Processing Co., Noro Base). The Department of Fisheries first approached the chiefs of Mbili in 1976 to see whether they would allow Solomon Taiyo to do test fishing on the Mbili grounds, and the chiefs agreed. Since then, the company has harvested bait fish from Mbili territory on a regular basis. Mbili is recognized as one of the most productive fishing grounds in the country.

When a boat is fishing for skipjack, the bait fish are released into the sea. Skipjack attracted to the bait are caught using pole and line. The bait fish themselves are harvested at night from shallow lagoon waters. First the small fish are attracted to the fishing boats by powerful electric torches dangled in the water. These torches are deployed both by the pole and line vessel and by a number of small companion boats (*esa*). *Esa* boats are each attached to a coastal village but are not necessarily crewed by local villagers.

After being concentrated by the lights, the fish are scooped from the sea using a net. Typically, boats take in an average range of 100-120 buckets (220-264 kg) of bait in one night, then head out to sea to fish for skipjack (Hansel Mamupio). Catches in Mbili are relatively high, being commonly

120-180 buckets per boat-night (Graph 3). Larger fish such as kingfish and barracuda are frequently caught in the net. This bycatch may be kept as food for the boat crew or given to villagers who come out to visit the pole and line boat. If the bycatch comes out of the water alive and unharmed, it may also simply be released.

The number of bait fish boats at work on the Mbili fishing ground varies from month to month and has also changed over the years (Graph 1,2). In the early days of the bait fishery there were as many as 22 pole and line vessels operating out of the Noro fishing base. In 1983, fishing boats spent 150 nights in Mbili, taking 17,912 buckets of fish. By 1988 this fishing had increased to 1076 boat-nights and 176,848 buckets of bait (Nichols 1988). Today a fishing boat may visit the bait ground only one to three times a month; in some months no boats come.

Pole and line tuna fishing vessel

Esa boat used to help capture bait fish

Owing to the ethnic tension that erupted in Solomon Islands in 1999-2000, the Japanese owners of Solomon Taiyo pulled their investments out of the country. They gave their share of the company to the Solomon Islands government, which now operates a reduced fleet of 12 pole and line vessels. Consequently, fishing activity on Mbili bait ground dropped dramatically (Graph 1). The Japanese crew were replaced by Solomon Islanders but none of them are from Mbili.

Impact on Bait Fish Stocks

Both the loss of particular species and the generally perceived reduction in bait fish stocks are often attributed to the industrial bait fishery that has been prosecuted on Mbili grounds for more than 20 years. Although locals perceive a decline in bait fish stocks, the fishing company maintains that fish are as plentiful as ever. The catch per boat-night at present is similar to that in the early days of the fishery (Graph 3). If stocks have indeed

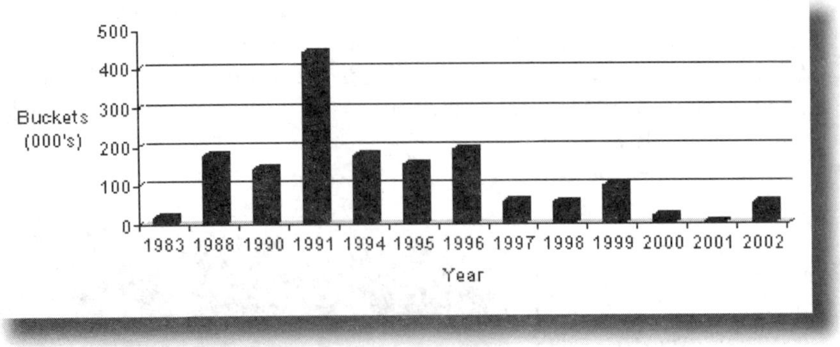

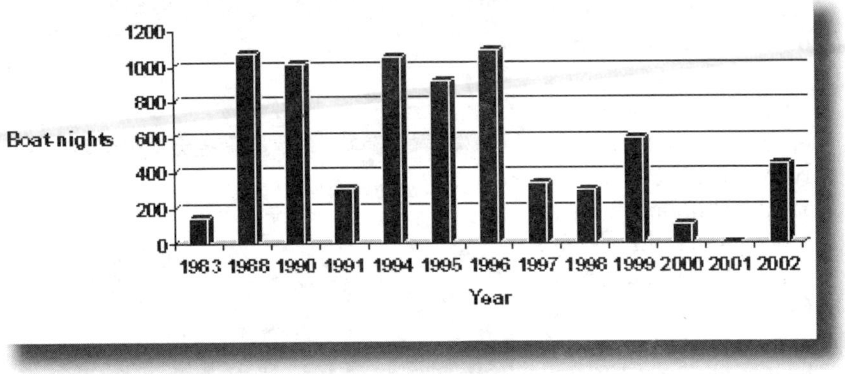

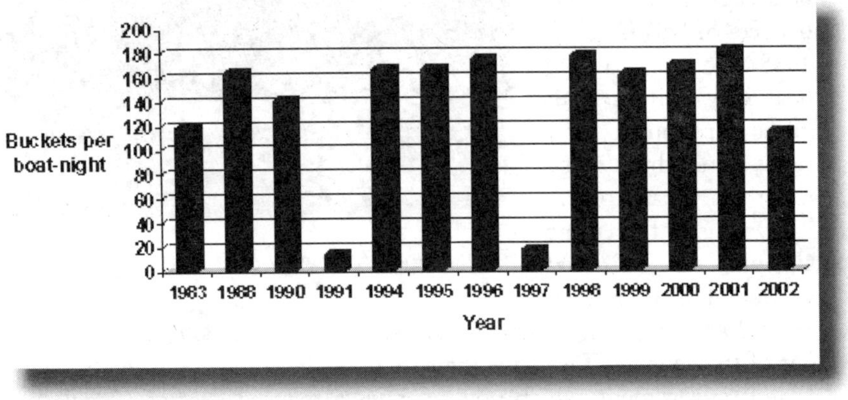

Graph 1. Thousands of buckets of bait taken from Mbili fishing grounds each year (note data missing for some years). Source of data: Solomon Islands Dept. Fisheries catch statistics.

Graph 2. Number of boat-nights that bait fish was taken from Mbili in each year.

Graph 3. Catch per unit effort (average number of buckets of bait fish caught per boat-night) on Mbili bait fishing grounds in each year.

declined then catch per unit effort may have been maintained through the use of increasingly efficient equipment, a concentration of effort on the best fishing nights in the month and/or increasing skill of the fishermen and fishing masters (John Legata, Mbili).

Blaber et al (1990 a, 1990b) stated that bait fisheries activity in the lagoon had no adverse impact on the coastal subsistence fisheries. On the other hand, Johannes et al (2000) and Hamilton (1999) maintained that Blaber did not have sufficient data to prove the point. Judging from these contradictory perceptions, more research is needed to settle the matter.

Economic Impacts

Although hundreds of tonnes of bait fish are extracted from Mbili fishing grounds annually, this fishery is of minor economic importance to the Mbili community. Mbili men do not work on the pole and line boats, and although several esa boats are based in Mbili, only one Mbili man has ever been employed on these small tenders, and then for only three years in the 1980s. Apart from some small trading opportunities and some fish provided to selected villagers by the boat crew, the main economic impact is from royalties that are paid to the chiefs by Solomon Taiyo Company in exchange for access to Mbili fishing grounds. Chiefs may also on occasion ask for and receive tuna from the company to supply a village feast or special event.

Mbili community members working on church construction under supervision of Chief Luten Watts (with his back to the camera)

After the initial negotiations with Solomon Taiyo the company agreed to pay a fee for each night a catcher boat visited the bait ground. On a good year such as 1988, up to SI \$35,508 (US \$4,793[1]) in royalties may have been paid to Mbili village chiefs. The chiefs take half for their personal expenses. They used to consult with village men on how to spend the rest of the money but this has not been done for many years. For the past 16 years the chiefs have been building a church using some of the royalty money but the church is still not finished. People faced with medical bills may ask for financial help from the royalty funds. When some people obtain money and others do not, discontent is sown in the community. Two of the three chiefs have removed themselves from the main village of Mbili, preferring to live in other settlements.

The Reef Owners Response to the Industrial Fishery

In 1975, in response to the increased activity of Solomon Taiyo Limited (STL) in the region, the reef owners around Marovo Lagoon came together under the leadership of a chief named Snyder Rini. The chiefs formed an association so they could collectively negotiate bait fishing rights with the company. The association tried to convince STL to pay royalties based on the number of bait buckets rather than a flat fee per boat night. Because catch recording is unreliable and there are no fisheries officers to verify the catch, STL refused to change. Since the government (Soltai) took over the tuna fleet they have also insisted on a nightly boat fee but whereas "previously the STL paid nightly flat rates of SI $27[1] for the smaller boats and SI $33 for the bigger boats… Soltai pay across the board SI $125 per boat per night" (Hon. Snyder Rini, Member of Parliament and chairman of the reef owners' association).

The reef owners' association is not well organized. There is no executive body and it is active only during negotiations at the end of each three-year fishing agreement period. Members do not always hold their ground during negotiations. For example in January 2001 when negotiations stalled some reef owners broke away and negotiated a separate agreement with the company (K. Assery, Noro base operation manager).

Social Impacts of the Industrial Fishery

Most village people do not want to discuss the royalties because they prefer to avoid conflict. In discussions and interviews performed for this study it was evident that, because they take community royalties for their personal use, the chiefs have lost some of the respect that would otherwise be accorded to them. Elders in the village express concern over this trend, which, in their minds, is part of the general disintegration of communal spirit in the village. They point out that in the past, when a chief or any elder called for assistance to build a house or engage in some activity for the general benefit of the village, everyone would turn out to help. These days, few will attend, and these will be only the person's immediate family.

In addition to the more generalized problems associated with the inequitable sharing of benefits, the bait fishery also has particular positive and negative impacts on women and children in the community. Mbili is a popular place where the pole and line crews go to shore to find food, play sports and socialize. The crewmen take up to 100 kg of their undersized skipjack with them and either exchange it for goods and services or give it away, most often to young women. These villagers in turn may consume the fish, give it away or sell it. Women are also invited out to visit the boat and to sell their vegetables and baked goods. Often they remain on board to socialize and they may be offered payment for sexual services. As a result, there have been pregnancies among ten young women over the years. These women have borne 15 children, 14 of whom are half Japanese. These children are integrated into the community, but their biological fathers provide little or no financial support. Most of the women have married locally and so have help in raising their children, but some remain unmarried and for them, life is a struggle. As one single mother explained "Now that the [Japanese] fathers of the children are no longer in Solomon they do not send money to help the child to school and so forth. Keeping the children is easy when they are small but now they are big it is very hard to keep them, because of the cost of school fees."

MANAGEMENT FOR THE FUTURE OF MBILI BAIT GROUNDS

This case study reveals the interplay of cultural, social, political, physical and biological factors that have affected Mbili fishing activities over time. The increasing population, entry into the cash economy and adoption of more efficient and destructive fishing technologies have combined to generate greatly increased pressure on marine resources. Increased fishing pressure and declining inshore resources have been reported widely in the Pacific (Sauni & Sauni, this volume) and it has been argued that access to pelagic species is an important factor that helps to conserve the more vulnerable reef stocks by providing an alternative source of protein (Aswani 1998). In Mbili this safety net appears to be compromised by the industrial bait fishing activities that compete directly with traditional inshore bait fishing. Competition between artisanal and industrial fishermen has been reported in other communities of Melanesia (Hviding 1996, Hviding & Baines 1994, Turner 1994). In the case of Mbili there are also social impacts (unplanned pregnancies and unsupported children) that are particularly hard on women and children.

In addition to an increase in population has been the increased participation of women and children, not just in subsistence fishing activities but also in commercial fisheries. Unlike in one case reported from Vanuatu (Tarisesei & Novaczek, this volume), women do not seem to have been denied access to modern technologies or to opportunities to fish commercially. This has been positive in terms of women's economic well-being but negative in terms of the increased pressure on fisheries. The increased numbers of people in the fishery and in particular the unprecedented involvement of women in commercial fishing activities has parallels in other Pacific communities (Vunisea 1997).

The church has had an impact on fish resources and on the social and political factors associated with the fishery. It was the establishment of the church that attracted people to settle in one coherent community, thus concentrating fishing effort. The church also prohibited certain kinds of fisheries. This is good in the sense that certain marine products remain untouched. On the other hand, it concentrates effort on the remaining fish species, which may then be more quickly reduced.

Destruction of marine habitat by land-based activities, climate change and political unrest that further stimulates the growing appetite for cash and foreign exchange, add to the factors discussed above and help to make the possible recovery of coastal fisheries uncertain.

What then is the future for the Mbili fishery? A recent reduction in pressure from the industrial bait fishery might allow at least the pelagic resources to recover, but there is no local institution to regulate fishing pressure or destructive technologies.

In many parts of Oceania (Hviding 1991) and also in eastern Indonesia (Novaczek et al 2001), inshore fishing grounds are controlled by traditional reef owners. The chiefs in some villages enforce gear type, size and seasonal restrictions on various fisheries and fishing grounds. Some also demand fees in return for access to particular commercial marine resources. In the province of Malaita, Solomon Islands, chiefs have historically placed taboos on reefs to control harvesting of shellfish (Fidali-Hickie & Whippy-Morris, this volume). Traditionally, chiefs in Marovo Lagoon also imposed gear and access restrictions on some reefs. This practice allowed fish to aggregate undisturbed and assured the community of a substantial catch for feast days (Hviding & Baines 1996). As noted by Aswani (1998), even where they do protect resources such practices are not necessarily designed with conservation in mind but rather serve to optimize foraging strategies. No such tradition was mentioned by Mbili people during community consultations but because the fishing grounds are "owned" by traditional leaders who have the power to restrict access to the resources, local management involving chiefly bans could be developed. Aswani (1998) has suggested that attempts to regulate fisheries should consider the pattern of individual fishing activities and fishermen's traditional knowledge. This is sound advice in the case of Mbili, where local knowledge of the place and time of spawning aggregations and migratory patterns of pelagic fishes is still held by community members. Aswani suggests that restrictions on fishing that mimic the seasonal patterns of fishers and that build on their local knowledge are more likely to be understood, accepted and therefore effective. Examples would be a ban on reef fishing in the season where most fishers are targeting pelagics, or the temporary closure of reef passages when fish are spawning. However, such strategies will only succeed if bait fish are sufficiently abundant to support local protein requirements, and if the physical environment has not changed to the extent that traditional knowledge has lost its value.

Social and political factors are also important when developing management strategies. When chiefly power is used for personal gain rather than in service to the broader community, it may be resented. If people lose respect for their traditional leaders they are less likely to comply to regulations and in fact may engage in environmentally destructive behavior as a form of political protest (Novaczek et al 2001:168, Tarisesei & Novaczek, this volume). Inequitable sharing of benefits from royalties in Mbili could likewise undermine local efforts in fisheries management and conservation.

In Mbili a major recipient of community royalty funds is the church. This is seen as good by the community members because individuals will not have to contribute to such community development from their own pockets and all members of the community will benefit because all will use the church. As the church still enjoys strong support as an institution and has historically paid attention to particular fisheries, it may be possible for the church to collaborate with or support chiefly authority in developing resource management strategies.

Finally, the increased participation of women and youth in the fishery must be recognized and accommodated. In Marovo Lagoon communities it is elder males, especially chiefs, who ultimately make the decisions over access to marine resources. In Marovo, unlike in some other traditional societies (Siamomua-Momoemausu, this volume, Tarisesei & Novaczek, this volume), it has been reported that women and young men have ways of exerting their influence on decision-making. Women and young men, because of their particular roles in the fishery, have important knowledge concerning current status of fish stocks and women in particular have been active in lobbying against certain forms of unsustainable land-based developments (Hviding 1991).

One further challenge to be met is the change in societal values that has come with modernization. How can one develop a management approach that builds on the strengths of tradition while accommodating the desires of younger generations for modernity and cash incomes? Similar issues are being confronted in other Pacific countries seeking to invent or reinvent community-based management institutions (Hickey & Johannes 2002, Veitayaki 1997). It has been suggested that an important aspect of reef tenure arrangements that should not be lost is the flexibility and adaptability of these traditional institutions (Hviding 1998).

The reef owners association offers a possibility for a lagoon-wide fisheries management body but it has never shown an inclination to protect and manage inshore reef resources. If the reef owners did decide to exercise control over inshore fisheries they would need scientific data and advice, obtaining which is another challenge for a country where research funding is scarce.

Experience elsewhere in the Pacific also suggests that intervention from an outside agency is sometimes needed to assist villagers to identify, prioritize and deal with complex resource issues. In the case of Mbili, there are several initiatives from outside of the community that might assist people to take charge of their situation. Under the leadership of SPREP (South Pacific Region Environment Program), an International Waters Project is scheduled to begin in Mbili in January 2004. This project will study the root causes of the degradation of coastal fisheries in Mbili and seek to strengthen community capacity for resource management and conservation (Kenneth Bulehite, project coordinator). Another outside initiative that may provide substantial benefits is the community forestry project implemented under the joint leadership of the University of Queensland and the Solomon Islands Rural Development Trust. This project will provide training in ecological forest management (Makin & Dart 2002) to revitalize coastal forests adjacent to the fisheries of Mbili.

For these outside initiatives to bear fruit, they must recognize and deal with local realities including the increasing importance of women and youth as stakeholders in the fishery. Documentation of local ecological knowledge and detailed accounts of current fishing activities can help to guide development of new management approaches. Issues of chiefly legitimacy, equity and the role of the church as an influential local institution, must also be addressed.

NOTE

1. At the time of writing SI $1 was worth US $0.135.

REFERENCES

Aswani, S.1998. Patterns of marine harvest effort in southwestern New Georgia, Solomon Islands: resource management or optimal foraging? *Ocean & Coastal Management* 40: 207-235.

Blaber S. J. M., D. A. Milton, N. J. F. Rawlinson 1990. Diets of the fishes of the Solomon Islands predators of tuna baitfish and trophic effects of baitfishing on the subsistence fishery. *Fisheries Research* 8:263-286.

Blaber S. J. M., D. A. Milton, G. Rawlinson, G. Tibora & P. V. Nichols 1990. Reef fish and fisheries in Solomon Islands and Maldives and their interaction with tuna bait fisheries. In S. J. M. Blaber & J. W Copland, *Tuna bait fish in the Indo-Pacific region: Proceedings of a workshop in Honiara, Solomon Islands.* Canberra: ACIAR Proceedings No.30.

Hamilton, R. 1999. *Tidal movements and lunar aggregating behaviours of Carangidae in Roviana Lagoon, Western Province, Solomon Islands.* MSc thesis, University of Otago, New Zealand.

Hickey, F. R. & R. E. Johannes 2002. Recent evolution of village-based marine resource management in Vanuatu. *SPC Traditional Marine Resource Management and Knowledge Information Bulletin* 14:8-21.

Hviding, E. 1991. Traditional institutions and their role in the contemporary coastal resource management in the Pacific islands. *Naga* (ICLARM) October 1991:3-6.

Hviding, E. 1996. Guardians of Marovo Lagoon: practice, place and politics in maritime Melanesia, Pacific Islands. *Pacific Islands Monograph Series* 14. Honolulu: University of Hawaii Press.

Hviding, E. 1998. Contextual flexibility: present status and future of customary marine tenure in Solomon Islands. *Ocean & Coastal Management* 40: 253-269.

Hviding, E.& G. B. K. Baines 1994. Community based fisheries management, tradition and challenges of development in Marovo. *Development and Change* 25:13-39.

Johannes, R. E., M. M. R. Freedman & R. J. Hamilton 2000. Ignore fishers' knowledge and miss the boat. *Fish & Fisheries* 2000(1):257-271.

Kupiainen, J. 2000. *Tradition, Trade and Woodcarving in Solomon Islands.* Finnish Anthropological Society TAFAS 45. Denmark:Intervention Press. P.184-197.

Makim, A. & P. Dart 2002. Environmental sustainability and poverty reduction: Pacific issues. *Development Bulletin* 58.

Nichols P. V. 1988. Bait fishery in Solomon Islands. In S. J. M. Blaber & J. W. Copland 1990, *Tuna Bait Fish in the Indo-Pacific Region: Proceedings of a Workshop in Honiara, Solomon Islands.* Canberra: ACIAR Proceedings No.30.

Novaczek, I., I. Harkes, J. Sopacua & M. Tatuhey 2001. *An Institutional Analysis of Sasi Laut in Maluku.* Manila: ICLARM,Tech. Rep. 59.

Otter, M. (ed) 2002. *Solomon Islands Human Development Report 2002: Building a Nation Volume 1.* Government of Solomon Islands Honiara.

Statistics Office 1997. Report 2: Village resources survey 1995/96. *Statistics Bulletin* No 10/97. Honiara: Ministry of Finance, Solomon Islands. 152p.

Statistics Office 1999. *Report on the 1999 Population and Housing Census.* Honiara: Ministry of Finance, Solomon Islands. 517p.

Turner, J. W. 1994. Sea change¨adapting customary marine tenure to commercial fishing. The case of Papua New Guinea's bait fishery. In G. R. South, D. Goulet, S.

Tuqiri, M. Church, *Traditional Marine Tenure and Sustainable Management of Marine Resources in Asia and the Pacific*. Suva: International Ocean Institute – South Pacific.

Veitayaki, J. 1997. Traditional marine resource management practices used in the Pacific Islands: an agenda for change. *Ocean & Coastal Management* 17(1):123-136.

Vunisea, A. 1997. Women's fishing participation in Fiji. *SPC Women-in-Fisheries Information Bulletin* 1:10-13.

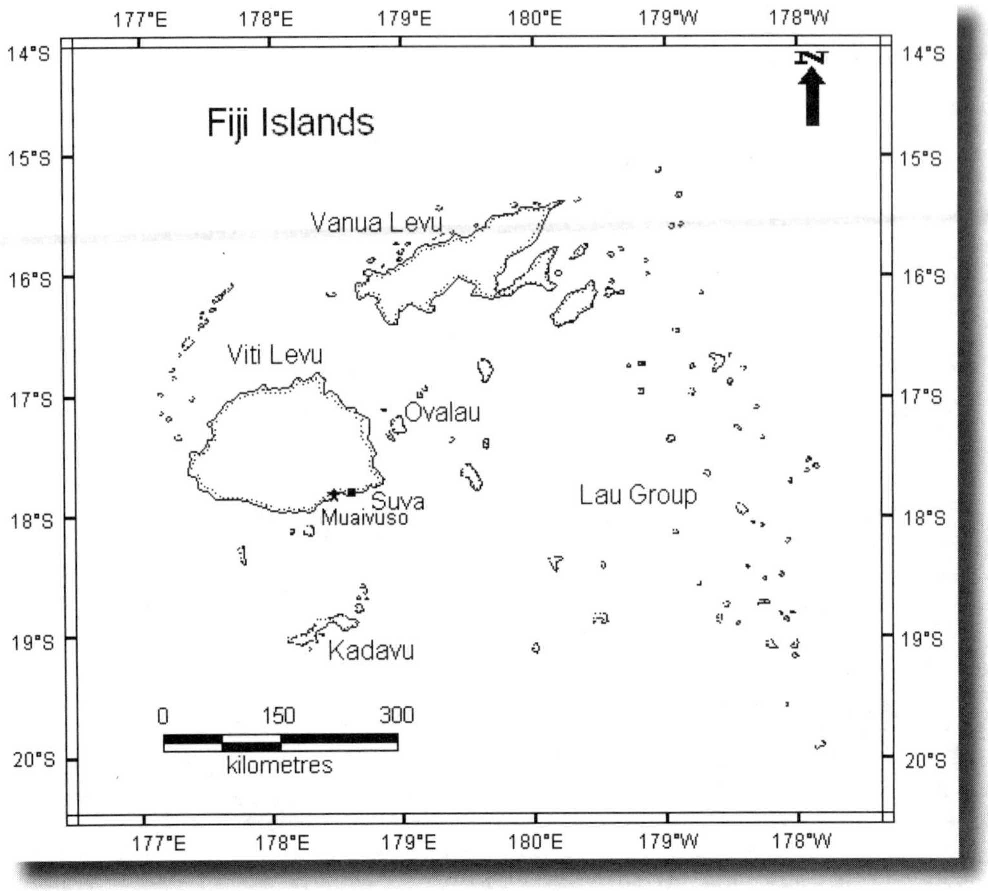

Map 1. Fiji, showing location of Muaivusu on the island of Viti Levu.

9 Is It Worth The Cost? The Live Rock Fishery at Muaivusu *Qoliqoli*, Fiji

Samasoni Sauni, Mecki Kronen, Aliti Vunisea, Lilian Fay-Sauni & Pierre Labrosse

"...what do you do when there is opportunity to make money in the village nowadays? In the past we never need monies as we enjoyed extracting resources from our seas. Now the tide has changed for we require having monies to meet the many family obligations. And that goes for even the least we thought was worthless, like live rock extraction..."

— *Ratu Rokola, Muaivusu.*

QUESTIONS FACING THE FIJI AQUARIUM TRADE

The aquarium trade, which includes the extraction of live aquarium fish, shellfish, living and dead corals and "live rock"[1] is relatively new and is fast growing not only in Fiji but all around the Pacific. The estimated value of the entire ornamental trade in the Pacific is not known but it is considered to be lucrative, generating large, short-term cash benefits and supporting national economies (Baquero 1999). What remains to be seen is the sustainability of such operations and the long-term effects on livelihoods of village people. This paper addresses some social and economic aspects of a live rock and live coral fishery in a rural village of Fiji.

People have, over generations, adopted and adapted fishing techniques and targets. Fishing people are well endowed with knowledge of their fishing grounds, so moving into a new fishery poses no particular difficulty. This is especially true for coral extraction, which requires no special skill and targets a previously unused resource. The problem with the trend to harvest live rocks and associated coral colonies is that coastal communities depend heavily on coral reefs because they provide habitat for fish, shells and other resources essential for food and income. Yet in most rural settings, coral reefs, even though they are critical to daily survival, are not recognized as having any real value.

Lovell & Tumuri (1999) outlined the main categories of coral extraction in Fiji: curio or ornamental coral, live aquarium products and live rock extraction. To date, there are ten companies collecting live reef organisms in Fiji. Muaivusu, a traditional Fijian village in proximity to the country's capital city, Suva, was selected as a site for investigating some socio-economic implications that may accrue from the development of a live rock fishery.

CASE STUDY METHODS

Socio-economic Survey

Fully structured, closed questionnaires were used to collect quantitative and qualitative data from households in Muaivusu. A general socio-economic survey carried out in August 2002 was followed up in May 2003 with surveys that focused specifically on live coral collection. Traditional and customary procedures were followed to inform the community, to receive permission to collect data and to implement each of the surveys.

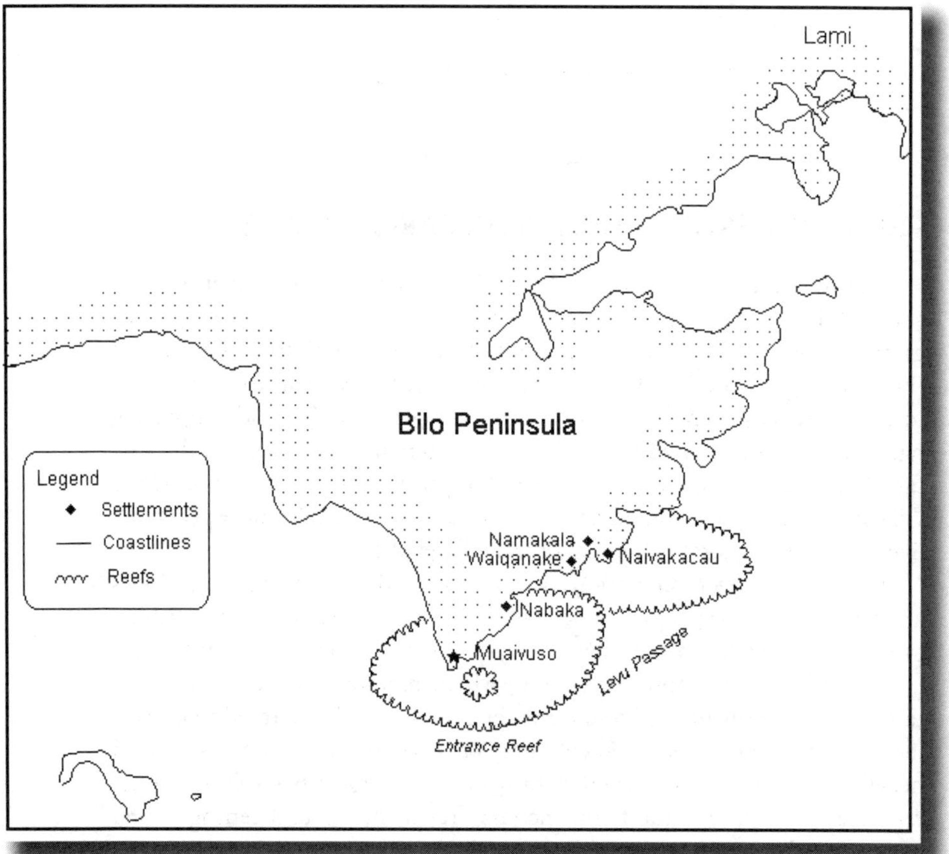

Map 2. Muaivusu and neighbouring villages.

MUAIVUSU DISTRICT

The study site (Map 2) comprises four villages having a total population of 1,273 people divided among 131 households, as follows: Muaivusu 36, Namakala 30, Nabaka 22 and Waiqanake 43. Households are large (average 9.7 persons per household). The traditional fishing ground *(qoliqoli)* covers 5 km^2 of fringing reefs and inshore lagoons. Two channels at either end of the reef enable easy, year round access to the outer edge of the reef under any weather conditions. The fringing reef is often exposed and accessible by walking during a low tide. Two lagoons (blue holes) are located behind the reef about 1 km from Muaivusu village and 1 km apart from each other. Branching corals are conspicuous on the silty-sand substrate along the fringe of both lagoons.

THE HISTORY OF LIVE ROCK EXTRACTION IN MUAIVUSU

Investigations revealed that the four villages in Muaivusu equally share ownership of their land and fishing grounds *(qoliqoli)*. However, only people from one of the four villages, namely Waiqanake, were involved in live coral extraction. Coral harvesting was carried out only during two distinct time periods.

The first harvest lasted for two years (1997-1999). Eight people from Waiqanake village extracted ca.10-20 crates of coral per day, twice to three times weekly. One of the eight fishers acted as middleman and took crates of coral directly to the factory which was based in Lami near Suva. A price of FJ $10 per crate was paid, rendering an income of about FJ $15-20 (US $8.70-11.60[2]) per fisher for each working day. The buyer provided a motorised boat and fuel. Only small pieces of live rock (predominantly massive coral coated with colourful coralline algae) were broken out of the reef. Fishers involved in the extraction reported that after two years, these particular corals were exhausted within the *qoliqoli* and so collection ceased.

The second coral harvesting project commenced early in 2000 and lasted until December of the same year. Based on a licensing contract between the Chief of Muaivusu and the Lami based agent, three experienced fishers from Waiqanake village were recruited to extract live coral. This time, the buyer provided only heavy-duty plastic buckets. Again fishers broke only colourful corals 15-20 cm in diameter from the reef. Collection took place regularly from 8 am to 5 pm on two consecutive days per week. The three fishers reported collecting 5-10 buckets per trip depending on weather conditions, earning FJ $13-27 per head[2] on each working day.

During both projects, fishers who were involved in coral extraction and their family members abandoned all other fisheries and spent their remaining time on farming. Among community members, attitudes toward the coral extraction were ambivalent. Both projects were possible only with the consent of the community's chief. The chief and the Department of Fisheries were responsible for making a choice between generating incomes on one hand, and contributing to the destruction of the community's *qoliqoli* on the other.

Entry into the Coral Trade

For the "officially recognised" harvesters, the activity represented easy money and the work could be shared with family members. "Everybody can collect live rocks and it is fun. We take the crowbars and get the rocks out, and my boys and me load them on to a bamboo raft. We sort the rocks on shore. The boys and also the girls can help us to load the rocks from the raft into the crates. Me and my fellow friends come afterwards and put the crates on the truck. Some of us drive to Lami where we give the rocks and crates to the agent. He pays us straight away....haaa, good easy money" (male, middle aged harvester).

As demonstrated by the quote above, coral or live rock extraction requires minimal skill and only brief training. Men, women and children from villages quickly pick up the trade. Training provided by the industry is usually in the form of identification of the rock and coral to be collected and the promotion of collection methods that ensure minimal damage to the product. Coral and live rock extraction are usually performed without any monitoring by the government or the companies and harvesters often have little regard for impacts of their harvesting activities on reef habitats.

Ecological Concerns

In 2000, a workshop was conducted in Muaivusu by the Fiji Locally Managed Marine Areas Network (FLMMA), in which destructive impacts of coral extracting were discussed. Fishers expressed concern about their previous involvement in live coral extraction. Nevertheless, the live rock industry was and still is considered potentially lucrative and attractive. As one woman explained, "They came here and held this workshop... the people from the University and the Fisheries and some other people, too. I don't remember them all, but they are part of what they call FLMMA. They told us all about the problems we cause when we collect the corals... the damage to the reef and losing the fish. So the chief and the community stopped all the coral-live rock extraction. But you see it gave good money, you know..." (older female fisher who harvests invertebrates and finfish in Muaivusu district).

Socio-economic implications

Interviews in 28 households in Muaivusu village provided data on aspects on demographic structure and size of households, the number of fishers and boats per household, household expenditures and the magnitude and sources of revenues. Analysis of the data revealed that the community was homogenous. There was no particular group within the community with a special interest in or aversion towards getting involved in a new fishery such as live rock extraction.

Table 1. Calculated annual catch and use of fish for Muaivusu qoliqoli.

Parameter estimated	Amount / value
Muaivusu total village catch	99 t/year
Catch per household that is fishing (Muaivusu, n=24)	4.125 t/year
Estimated total catch for all fishing households in district (n=111)	458 t/year
Muaivusu village per capita consumption of fish	105.7 kg/year
Total population of district	1273 people
Estimated total consumption of fish in Muaivusu district	135 t/year
Balance of fish available for sale	323 t/year
Average price per kg reef and lagoon fish	FJ $3
Total value of fish exported per year	FJ $969,000
Mean household income from fish sold (111 households)	FJ $8,730/year

Following this general characterization of the village we tested for what overall income variations might arise if people adopted coral extraction instead of marketing traditional fishery products (Table 1). Fisheries data collected at Muaivusu village was extrapolated to estimate a total for the district. Taking into account that 20 households were not engaged in any fisheries at all, the remaining 111 households of the district account for a total annual catch of 458 tonnes of fish. Most of this catch (71%) is sold. Total value of the catch sold is estimated on the basis of an average price for fresh fish of FJ $3 kg^{-1}. Thus, total cash income from fish for the district is estimated at FJ $969,000 per year. Averaged over the total number of households that fish, this represents a mean annual income of FJ $8,730 (US $5,063) for each of the 111 households.

Taking into account that each Muaivusu village household has on average 1.4 fishers, we assume a total of 155 fishers in the Muaivusu district. On the basis of four working days per week, this equals an average daily income of FJ $8 for each reef fisher, or $32/wk (US $18.56[2]).

Under good weather conditions a daily income of FJ $15-20 per fisher was reported for extracting coral. However, coral extraction was performed only two or three days per week. Thus, the weekly cash income from the coral harvesting was in the range of FJ $30-60/wk (US $17.40-34.80). This is up to FJ $28 more than the average weekly income from reef and lagoon fisheries. In addition, live rock/coral harvesters gained one or two days that they could invest in any other activity, such as gardening.

One indirect impact of live rock/coral extraction is that people who collect live rock stop other fishing activities and therefore forgo not only an income of FJ $8,730 per annum, but also their food fish. At least part of the added income from coral harvesting or gardening would therefore have to be used to buy meat that was previously gained through fishing. Because of the withdrawal of coral harvesters from fishing, the overall finfish catch of the community would decline and some of the remaining catch would have to be diverted to cover these families' needs for fresh fish to eat.

Data from Muaivusu village indicated that on average, each household member was consuming 105.7 kg fish per year (Table 1). If this had to be purchased it would cost FJ $317, or FJ $6 per person per week. The weekly cost to provide fish for an average household of 9.7 people works out to FJ $58. The family income from coral harvesting was at best FJ $60/wk. Therefore, unless fish could be obtained free from the extended family or traded for excess garden produce, a large fishing family could suffer economically from choosing to stop fishing and take up coral harvesting. If these families obtained fish for food from relatives in the community without the need for cash payment (a likely scenario given the local culture), this would leave the cash benefits in the hands of coral harvesters but would share the costs across the extended family. Ultimately the result would be to either reduce the community's overall income from sales of fish or cause some people to reduce their consumption of fresh fish in order to maintain their cash incomes.

THE FUTURE OF THE LIVE ROCK – CORAL TRADE

Socio-economic data showed that a few households enjoyed short-term cash benefits from coral harvesting but, because they abandoned finfishing, they had to find some other way to obtain fish to eat. For a large household this could eliminate the economic benefits of coral harvesting. The small number of people involved coupled with the short-term period of their operation meant that, at a community level, there was little distinguishable positive or negative economic effect of the live rock fishery. Depending on whether or not coral harvesters were given fish to eat for free, the overall impact might be a small decline, either in community income from fishing or in fish consumption.

The potential daily income gained by the officially recognized, adult harvesters of live rock is slightly higher than that from traditional fishing. Remuneration depends on the amount harvested and its quality. Although women and male and female children may assist with harvesting, it is the key (generally male) harvester who gets paid. This type of arrangement could result in a relative disempowerment of female family members, compared to fishing activities that involve both genders but for which females are the predominant salespersons.

Athough it is a cause for concern among those aware of its potential impacts on the environment, the live rock fishery is still considered an economically attractive option compared to other village activities. Perhaps a key argument in its favor is that sales of coral are guaranteed and money is paid immediately. By comparison, the catch from traditional fisheries may be in part consumed as a result of social obligations. Also, fishing families must spend time, money and energy transporting the fish to market. Conditions in markets are such that fish may spoil or go unsold, meaning that income is not assured.

Coral reefs and reef-associated organisms have provided basic needs to coastal communities for centuries. The extraction of coral using metal implements such as crowbars is clearly destructive. Reef habitats are directly disturbed both by the rock extraction and when rafts loaded with rock are dragged ashore. A key question, which is not dealt with in this case study but reported elsewhere, involves the response of reef fish communities to coral extracting, and whether and at what rate they can recover. There is evidence from biological research that at Muaivusu, the numbers of some reef fish are significantly fewer over areas of reef where coral harvesting took place (Sauni et al unpublished). The socio-economic benefits derived from such operations must therefore be weighed against the disruption of the reef ecosystem.

The growing live rock fishery and the aquarium trade present both opportunities and challenges. There is a need to put in place management and monitoring strategies to ensure that the trade does not have an impact on the resource owners' livelihoods in the long term. Also, although people in rural coastal communities need more diverse economic opportunities, these should not create a "culture of dependence" where natural resources that sustain livelihoods are increasingly valued only in monetary terms. That said, it is true that in some communities, engagement in the coral trade has helped to build basic amenities, and has brought much-needed economic development. In the absence of comprehensive data and analysis, villagers, developers and managers face difficulties in deciding between the potential environmental impacts of the trade (which may be largely hidden and difficult to quantify) and the potential cash benefits (which may seem more obvious but may not be entirely real).

Currently, live rock extraction activities are concentrated in a limited number of areas in Fiji. Where this fishery is performed at a larger scale and over a longer time period than in the case of Muaivusu, coastal communities may indeed earn significant amounts of money. However, there is a real concern that there will be long-term impacts from the extensive collecting of live rock and coral within these fishing areas. The loss of productivity of fishing areas, possibly on a long-term basis, may be significant. Also, as shown in this case study, a closer look at economic costs and benefits may well reveal that for some families economic benefits are an illusion or costs are borne by the larger community without being recognized. The Muaivusu case also reveals that a more detailed gender analysis is needed to see whether there is equitable sharing of benefits and burdens by male and female family members.

This case study highlights the urgent need to address the general lack of understanding of the ecological effects of live coral/rock extraction (Hughes & Connell 1999). As a precautionary measure, countries in the region should impose regulatory measures on this particular aquarium trade. There is also a need for large-scale studies to gain a more thorough understanding of the socio-economic effects of the coral and live rock fishery in Fiji, and elsewhere in the South Pacific. Muaivusu may serve as a good example where resource-user adaptation (abandonment of fishing effort after colorful corals in the fishing ground were exhausted after two years) and increased environmental awareness (the result of community based management interventions) combined to prevent long-term and measurable damages. In communities where more intensive and long term extracting is undertaken, there is potential for greater impacts leading to unbalanced ecological and socio-economic structures.

NOTES

1. Live rock is rubble or reef rock covered or permeated with living coralline algae and other marine life.
2. At the time of writing (2003), FJ $1 was equivalent to US $ 0.58.

REFERENCES

Baquero, J. 1999. *Marine Ornamentals Trade – Quality and Sustainability for the Pacific Region.* Suva: South Pacific Forum Secretariat and the Marine Aquarium Council. 73 p.

Hughes, T. P. & J. H. Connell, 1999. Multiple stressors on coral reefs: a long-term perspective. *Limnology and Oceanography*, 44:932-940.

Lovell, E. R. and M. Tumuri 1999. *Provisional Environmental Impact Assessment for the Extraction of Coral Reef Products for the Marine Aquarium and Curio Trade in Fiji.* Report for Fisheries Division, Government of Fiji, June 1999.116 p.

Sauni, S., M. Kronen, A. Vunisea, L. F. Sauni, R. D. Vave, P. Labrosse (in progress). Assessing the live rock fishery at Muaivusu *qoliqoli. South Pacific Journal of Natural Sciences*, USP, Suva.

10 *Trochus* Resources: A New Fishery For Tonga?

Siola'a Malimali

"Tongan households have traditionally relied greatly on inshore fisheries resources to meet subsistence needs. Around one third of Kolonga households still engage in fishing for subsistence or cash purposes."

— *Sione Vao, Kolonga village fisherman, age 76*

"I never heard of the Ministry of Fisheries program and I do collect Trochus as part of my fishing catch... I collect everything good for eating... Trochus has a big meat compared to other snails, which is good."

— *'Elitisi Filikitonga, Kolonga village fisherwoman, age 62.*

INTRODUCTION

As pressure on nearshore resources increases and indigenous fisheries decline, many Pacific Island countries have begun to explore the potential of aquaculture for enhancing local food security and providing marine products for export. In some cases, countries have embarked on what in many parts of the world would be considered a high risk gamble. Exotic species are liberated into fish habitats to grow and spread, in hopes of developing to the point where they can be commercially fished. Thus, *Tilapia* fish have been and are still being released into rivers in Fiji, with dire consequences to small indigenous fishes that may be out-competed and driven to the edge of local extinction (Aaron Jenkins, Wetlands International, pers comm 2004). In the sea, the most popular species for introduction has been the topshell, *Trochus niloticus*. This species reportedly out-competes at least one indigenous gastropod, the green snail (Sims 1984). Tonga embarked on the experimental release of *Trochus* in 1992. Even though efforts have been directed to the study of *Trochus* transplantation and of hatchery-produced *Trochus* in Tonga (Manu et al 1994), information on the size of the stock and fishermen's perceptions are limited. This case study is an initial attempt to evaluate some of the social consequences of the introduction.

The focus of the case study is the village of Kolonga on Tongatapu. Kolonga fishers enjoy collecting *Trochus* and seem to have no idea at all that it has been introduced, that harvesting of *Trochus* is banned, or that a stock enhancement program is being run by the Ministry of Fisheries. The need to work on the biological aspects of *Trochus* introduction and conservation has been recognized and acted upon by the government, but fishers' activities and attitudes have been overlooked. These need to be understood in order to establish a *Trochus* fishery. In this case study, I will explore how fishers' activities, attitudes and livelihood options influence the *Trochus* fishery establishment program. I will also consider what might be an appropri-

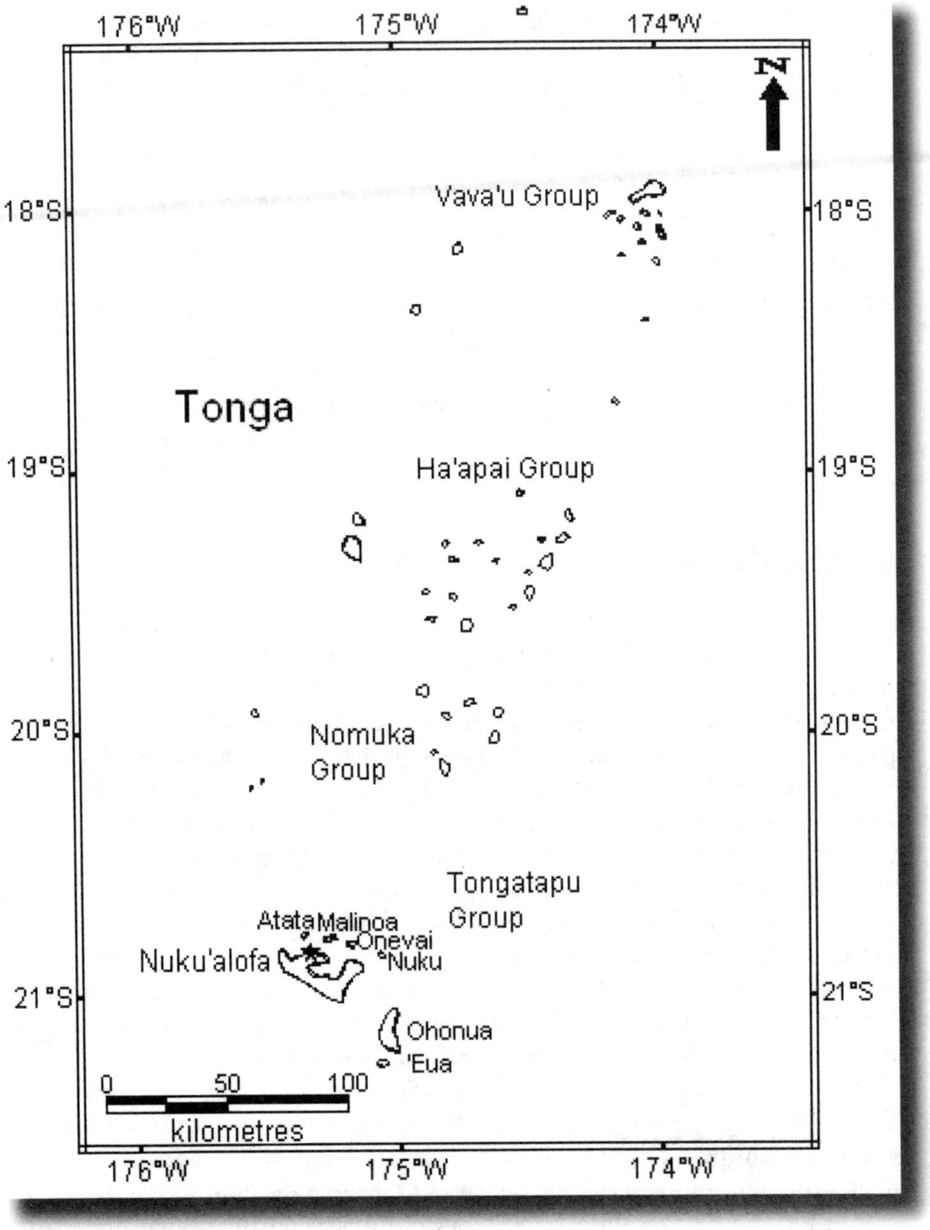

Map 1. Tonga, showing the study site on Tongatapu.

ate approach to promote the establishment of a well-managed and viable *Trochus* fishery. As is often the case when dealing with shallow inshore resources, the role of women in the fishery is revealed as a critical element for consideration.

SUBSISTENCE FISHERIES AND FISH HABITATS IN TONGA

The Kingdom of Tonga comprises 170 scattered islands (Map 1) of which 36 are inhabited. There are three main groups of islands; the estimated total land area is 747km^2. The coralline and limestone islands in the Tongatapu, Ha'apai and Vava'u groups are immediately west of the Tonga Trench, while further west is a line of small volcanic islands, some of which are still active. The Exclusive Economic Zone (EEZ) of Tonga as defined under the Law of the Sea covers an area of approximately 700,000km^2, much more than the 395,000km^2 area claimed under the 1887 Royal Proclamation (Bell et al 1994).

This case study deals specifically with the inshore fishing grounds of Tongatapu Island, which have been estimated to cover 947 km^2. Reefs and mangroves make up 11.2 and 0.36 per cent of the area, respectively. The remainder consists of lagoons and an area of outer shelf that is less than 160m deep.

Fishing has always been an important subsistence activity in Tonga, with the lagoons and shallow-water reefs providing vital protein for the local population. With the change from barter to a cash economy, these resources have been subjected to increasing pressures. There is evidence that inshore resources are over-fished. Zann (1994) stated that Tonga's inshore and deep-water fisheries were moderately to seriously overexploited; catches of most reef species have been gradually declining.

The basic fisheries law in Tonga is the Fisheries Act of 1989, which provides for the management and development of fisheries. The new Fisheries Management Act of 2002 provides for the conservation, management, sustainable utilization and development of fisheries resources in the Kingdom and its EEZ. This Act applies limits to the access to fisheries resources.

Although fisheries are often portrayed as male activities, women play a major role in subsistence fisheries that provide seafood for communities in Tonga. The degree to which women are involved in fisheries varies from place to place but the main aim of women's fishing is always to provide food for the family. In addition, women may collect seafood to be sold, in which case the money is used for family expenses.

Women have always been involved in the exploitation of fisheries resources. They are not forbidden to use modern fishing methods but, perhaps because of the high status of women in Tongan society and the cultural emphasis on feminine beauty, they restrict their involvement in fishing to activities that are close to home and not too strenuous. Women typically glean shellfish, seaweed, jellyfish, sea cucumber and sea urchins on the reefs. Men carry out the more physically challenging fishing activities such as diving in deep water and hauling nets.

TROCHUS SHELL FISHERIES OF THE SOUTH PACIFIC

The History of *Trochus* Introductions

Trochus shell, also known as topshell *(Trochus niloticus)*, has been one of the most widely exploited gastropods in the South Pacific in recent decades. *Trochus* live on coral reefs from the inter-tidal zone down to about 15 metres depth (Gillett 1995). Although *Trochus* is used primarily in the manufacture of valuable mother-of-pearl buttons, other minor uses include jewellery and artwork. In addition, the meat is edible and it may find international markets (Sims 1984). Bour (1992 in Manu et al 1994) reviews the status of the *Trochus* resource in the region.

Pile of Trochus shells taken from the Kolonga reef.

Trochus occurs widely in the Indo-Pacific region, but it is not native to some of the countries where it is currently found. The natural geographical distribution of *Trochus niloticus* extends only from Wallis Island in the central Pacific westward to Sri Lanka and from southern Japan southward to New Caledonia and northern Australia (Nash 1993). In the 1920s and 1930s, *Trochus* was transplanted from Palau to the Micronesian Islands. This transplantation was remarkably successful in some areas. In the 1950s, transplantation began in the South Pacific region, as reported by Gillett (1995). *Trochus* has now been successfully introduced into a number of countries in the Pacific region, where it constitutes an important source of revenue and protein (Parkinson 1984). For example, in the Cook Islands, where *Trochus* was introduced in 1957, it has been fished commercially since 1981 and is also an important subsistence fishery. *Trochus* has become a substitute for traditional reef foods that are no longer abundant in the Cook Islands, thus releasing pressure on resources that may otherwise have been harvested to near extinction (Ponia 2000).

Gillett (1995) reported that the majority of the world's *Trochus* production for the 1990s (59%) was from the Pacific region. *Trochus* production during the period 1985-1994 was about 2,300 metric tons annually. The export value of this production is estimated at US$15 million per annum (ICECON 1997). Not surprisingly, there has been considerable interest in Tonga in pursuing similar benefits based on a successful transplantation program.

The Introduction of *Trochus* into Tonga

The history of *Trochus* introduction into Tonga has been one of success (Gillett 1993, Fa'anunu 2001). It began in August 1992 when the first batch of *Trochus* was air freighted from Fiji (Gillett 1992). There followed two introductions in 1994 and 1995 (Niumeitolu et al 1999, Gillett 1995). The primary goal was to create new commercial opportunities for the people of Tonga. As a consequence of these introductions, *Trochus* are now found on reefs close to their release sites (Niumeitolu et al 1999). The Tonga Department of Fisheries has developed hatchery techniques and hatchery-raised juveniles have been released onto selected reefs around Tongatapu to accelerate the establishment of *Trochus* stocks. Harvesting of *Trochus* is officially banned in Tongan waters. The potential for a future fishery is still unknown.

METHODOLOGY

Case study research started with study site selection, for which geographical position and fishing activities were considered. The chosen site for this study, Kolonga, is one among a number of coastal villages whose fringing reefs are close to the *Trochus* release sites on the islands of Nuku and Fukave (Map 2). Kolonga's reefs are a barrier to the south of the release sites and therefore are areas where drifting *Trochus* larvae may settle. The release sites are major fishing grounds for these villagers. It was therefore suspected that Kolonga fishermen might currently be fishing *Trochus*, despite the harvest ban declared by the Ministry of Fisheries.

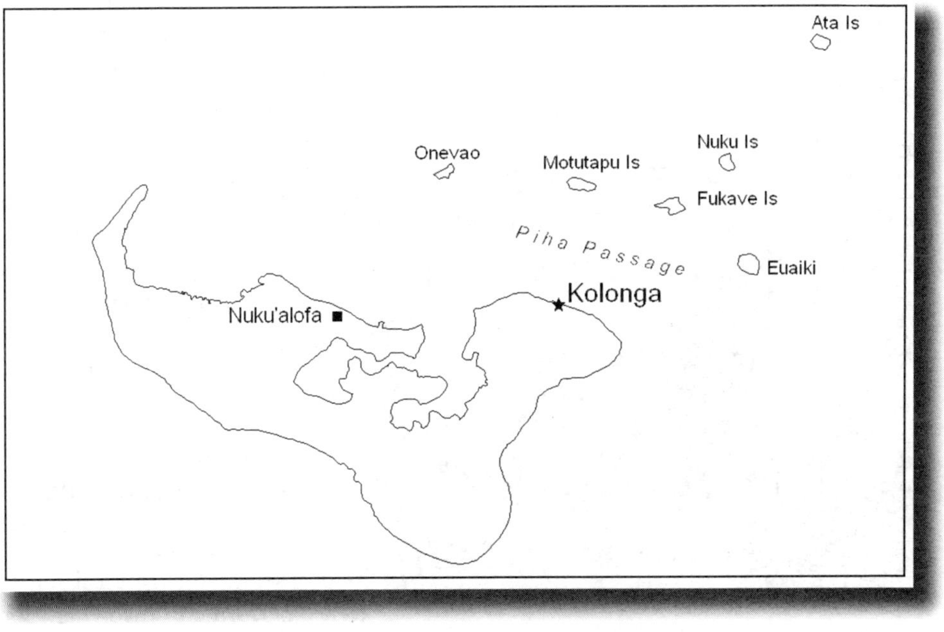

Kolonga village and fishing area, showing the Trochus release sites on Nuku and Fukave.

Observation of fishing activities was carried out at Kolonga for four consecutive days. From 20 October to 16 November 2003, surveys of fishers on the fringing reef at Kolonga village were carried out using structured questionnaires. For 14 days, individual fishers' catches were inspected (creel survey) during or at the end of their fishing trips. *Trochus* found in fishers' catches were counted and measured and the percentage of *Trochus* in shellfish catches was estimated.

Some fishers were selected as key informants on the basis of whether they had *Trochus* in their catch or admitted to being active *Trochus* fishers. Key informant interviews were carried out over a period of five days. In addition to fishers, several other people were asked for specific information, including the Kolonga Town Officer.

KOLONGA VILLAGE AND ITS FISHERIES

The Village

The population of Kolonga is about 1700, distributed over 315 households (Town Officer pers comm). The major daily activities in the village are farming and fishing, which are also the main sources of income. The land is owned by a chief, but the people have free access to the sea for fishing.

The Kolonga fishing ground is a fringing reef adjacent to the village. People gain access to the fishing ground by foot, canoe and motorized boat. Fishers include both men and women, ranging in age from their twenties to sixties. The types of fishing practi ed in this area include day and night diving, day netting (gill net and cast net) and hand lining, and day and night gleaning.

A woman who gleans shellfish and octopus on the Kolonga fishing ground carries her catch in a hand-woven basket.

The back reef area of Kolonga's fishing ground is dominated by seagrass and rubble patches whereas the reef flat to reef slope areas are dominated by patches of living hard coral. Observation of the reef during the survey confirmed that it has been over fished. There were hardly any living organisms to be seen. Fishers reported that their catches were very small compared to catches in the past, and consisted mostly of shellfish.

Fishers and Fishing Areas

In the course of the 18 days of field observations, 76 creel surveys were performed involving four different groups of fishers: day fishers (mostly gleaners), night gleaners, day divers and night divers (Graph 1). Most of those fishing on the Kolonga reef (96%) were from Kolonga; a few resided in neighbouring villages. Gleaners said they fished only on this reef or, if they were from another village, then also in front of that village. Day and night divers indicated that they sometimes swam in front of several villages when fishing. Night divers mostly went out to the small islands of Nuku and Fukave where *Trochus* were originally released. Sometimes they travelled farther away to Eueiki (Map 2).

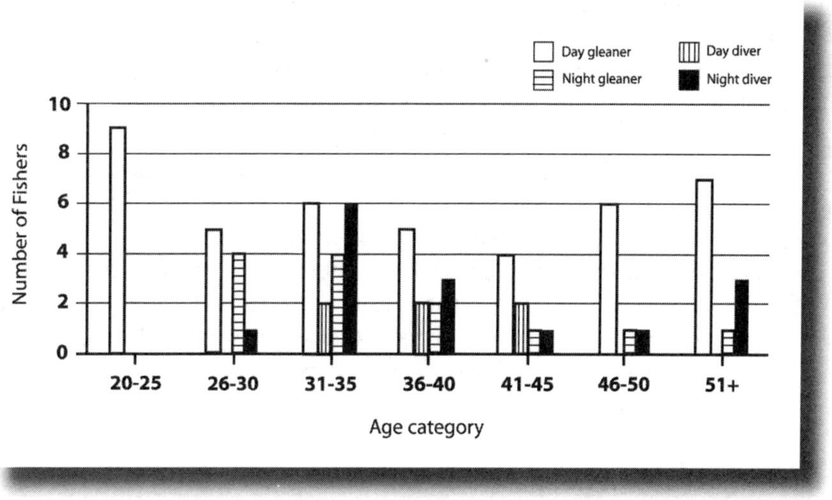

Graph 1: Type and age distribution of Kolonga fishers

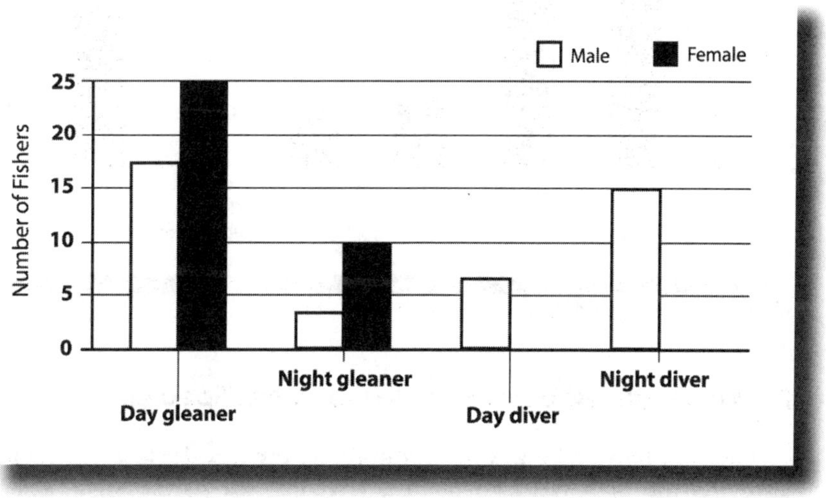

Graph 2. Gender distribution of Kolonga fishers

Day gleaners were the numerically dominant fishers in all age categories (Graph 1) and the majority of gleaners were female (Graph 2). Thus, day gleaning by women is the major fishing activity influencing the Kolonga reef. All female fishers encountered were gleaners and used only their hands or a metal spear to capture their prey. A minority of gleaners were male. Looking at the age distribution of gleaners, it is clear that most are in the youngest and oldest age groups. In other words, gleaning is an activity that is important both to young people under age 25 and to elders over 50. The youngest gleaners are almost all female whereas the oldest are more evenly divided among males and females (Graph 1).

All of the divers found working Kolonga reef were male (Graph 2). The majority of these male fishers were 30-40 years old.

The most common source of income of the fishers was fishing, followed closely by farming (Graph 3). All the fishers who fished for cash income were men. Female gleaners fished only for food. These women gained income mostly from farming, but also from handicrafts and remittances. Fourteen of the 16 gleaners who reported no cash income were female. The majority of the female fishers described their vocation as "housewife".

Day gleaners dominated the group of most intensive fishers, who fished 5-6 times a week (Graph 4). These fishers appear to be heavily dependent on gleaning for subsistence. Most other gleaners were also very active, fishing 3-4 times a week. Most of the divers, who were male, also fished 3-4 times a week.

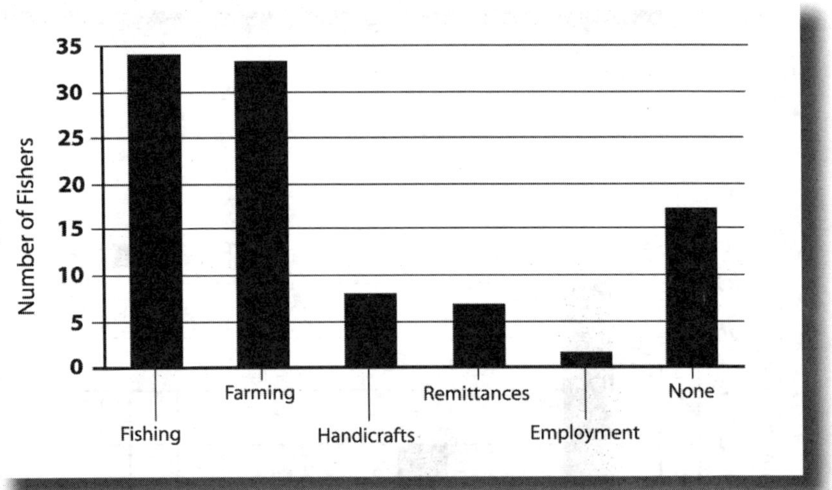

Graph 3: Main sources of income for Kolonga reef fishers.

Fishing Gears, Habitats Fished and Target Species

Gleaners (day and night) work mostly on the reef flat and crest; some also fish the back-reef/lagoon area, targeting sea hares. Night gleaners primarily target sea hares but three people said they pick everything edible and others included finfish and shellfish as secondary targets. Day

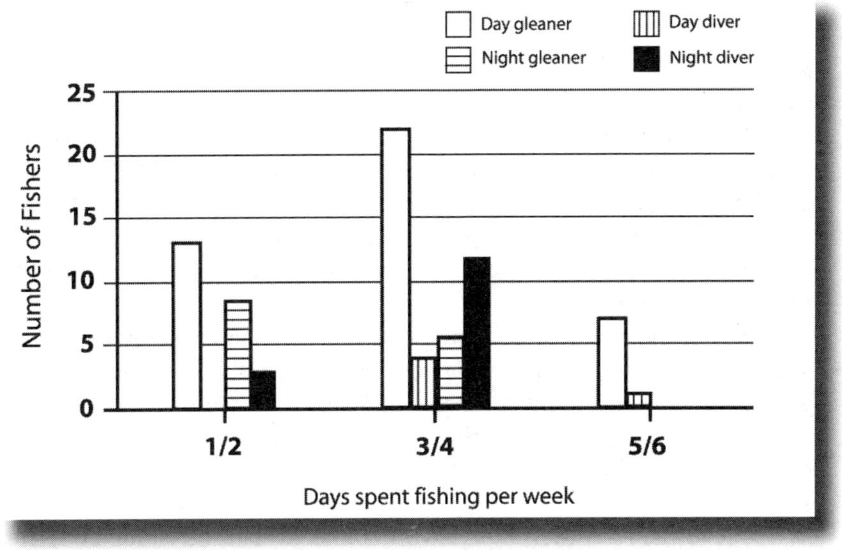

Graph 4. Intensity of fishing (fishing days per week) for each type of fisher on the reef.

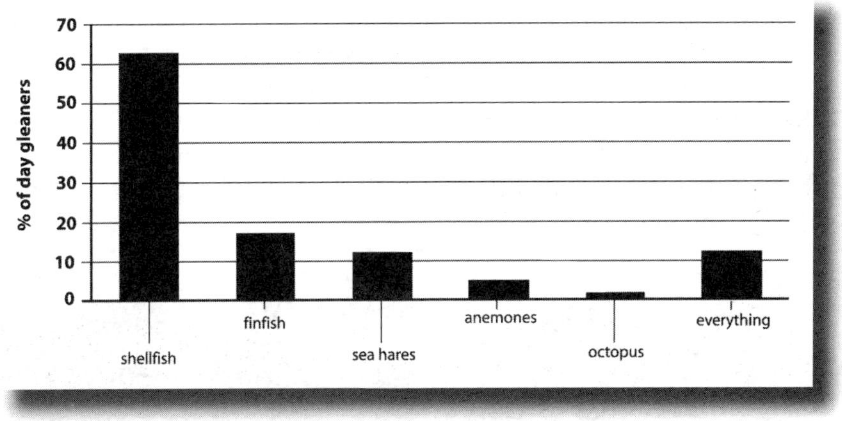

Graph 5. Percentage of day gleaners targeting each of the primary fishing targets

gleaners have a wider range of primary targets (Graph 5) but are mostly after shellfish. Other minor targets include sea cucumbers, sea urchins and seaweed. All night gleaners and most day fishers harvest by hand but four day fishers were using hand lines; two used a gillnet or cast net. All of these day fishers who used nets or lines were male.

Day divers work on the reef slope and their main target is octopus. They may also pick up shellfish, finfish and sea urchins. Night divers typically take boats over to small offshore islands and dive on reef slopes there. Night divers target finfish almost exclusively but may also collect lobster.

TROCHUS HARVESTING

Trochus were found in the catch of day gleaners and day divers, but not in the catch of night fishers. Only eight out of the 79 catches inspected included *Trochus*. Three of the fishers with *Trochus* were female gleaners, three were male gleaners and two were male divers. All picked *Trochus* for home consumption; none sold shell, nor did they know that the shell was marketable. On average only 3% of the shells in their creels were *Trochus*. Compared to other shells, *Trochus* are large and easy to collect. The small percentage of *Trochus* catch therefore reflects the limited abundance of these large shells. It is important for fishers to understand that although the number of shells they are collecting may seem insignificant, the capture of even small numbers makes the chance of establishing a *Trochus* fishery in the future uncertain.

Many of those fishers who had no *Trochus* in their creel reported seeing and sometimes collecting *Trochus* on the reef crest and slope. Most day gleaners (71%) and night gleaners (85%) and all day divers interviewed collect *Trochus* when they see them. None of the night divers interviewed collect *Trochus* although 53% of them have seen the shells on the reef slope.

None of the 76 fishers interviewed had heard that the taking of *Trochus* was not allowed. A few said they thought that they had been catching *Trochus* since 1998 or 1999; most could remember seeing *Trochus* only in the past few years, since 2002. Harvesting is easy whether by diving or gleaning, and the *Trochus* meat is considered tasty and desirable.

The allowable harvest size in other countries with a fishery, for example Solomon Islands, is greater than 6cm (Foale 1998) because the sexually mature size of both male and female *Trochus* is between 5.5cm and 7cm basal diameter. The minimum size limit used in Vanuatu is 9cm (Amos 1995). This limit has been set so that *Trochus* can breed a sufficient number

Kolonga fisherman who unwittingly fishes Trochus, with Trochus shells.

of times to maintain the population before being harvested. Where *Trochus* is harvested for the shell, the oldest, largest shells are often left behind because they are too damaged to be saleable. This is good for the stocks because the larger a *Trochus* is, the more larvae it can produce.

In Kolonga the *Trochus* harvested ranged from 30mm to 120mm, with the majority being 50-90mm (Graph 6). This means that some of them didn't have a chance to reproduce and contribute to the establishment of a *Trochus* fishery in Tonga. If these larger animals continue to be fished, and especially if even the largest, damaged shells are taken for meat, it will reduce the number of spawners available to produce larvae. As a consequence, a longer time will be needed to establish a harvestable stock. Because it is small and vulnerable to fishing, the *Trochus* population could even be depleted and go locally extinct.

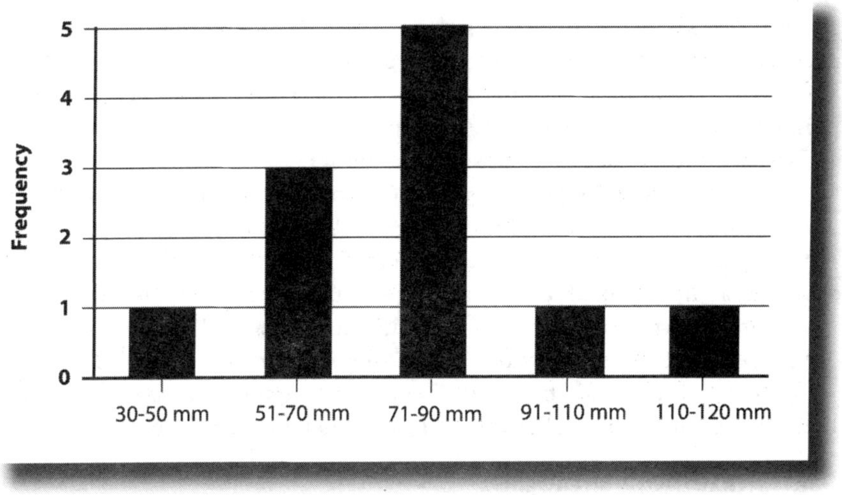

Graph 6. Basal diameter of Trochus harvested from Kolonga reef.

THE IMPORTANCE OF *TROCHUS* TO SUBSISTENCE FISHERS

To interpret this field survey, it is important to be aware of the livelihood of the people of Kolonga. Many households depend on inshore resources for subsistence food needs. Because the fishing ground has been over fished, fishers collect anything that is edible. Under Tongan law, all people are free to fish wherever they wish under what is known as an open access system.

It is evident from inspections of fishermen's catches that significant numbers of *Trochus* are now fished on reefs such as Kolonga, which are at least 5km distant from the original release sites (Fa'anunu unpublished). There is clearly recruitment of *Trochus* to reefs around Tongatapu and the large size of some of the individuals taken by fishermen indicates very good growth rates for such recruits. The transplantation of *Trochus* has proved to be a biological success.

Despite the success of the transplantation program, the primary objective is to allow a new fishery to be established, *Trochus* stock may not reach a commercial level. This study shows that fishers have already started to collect *Trochus* for food; 3% of the shellfish catches assessed were *Trochus*. *Trochus* can be collected easily, requires no costly equipment and the meat is prized. The reef areas in which *Trochus* are mainly found are shallow reef flats that are often emergent at low tide, as well as the first few meters of the reef slope. It is this distribution that determines the ease of fishing by hand collection on foot (gleaning) or by free diving.

The female gleaners who were surveyed do not sell their catch and many have no source of cash income. This does not mean that there is no market for *Trochus* meat. It is because the family income is often provided by a spouse or from other activities such as farming and handicrafts. Nevertheless it is likely that many women depend heavily on gleaning, and especially on shellfish, for family protein, particularly in families where incomes are limited. In addition to filling this basic food requirement, people engage in fishing because it is a lifestyle choice. At low tide, many people go out fishing for recreation; it is part of their culture.

Shellfish are an important source of food, as indicated by their prominence among the primary fishing targets. The attitude of Kolonga's fishers toward *Trochus* resources is clearly reflected in the results of the survey. The fishers do not think of *Trochus* as an introduced species that is being reared for future purposes. They treat it just like the other shellfish that are fished for food. The people may be consciously violating the fishing ban of the Ministry of Fisheries but they seem to be genuinely uninformed. This results in continuous collecting of *Trochus* shell despite the ban.

Management strategies

Fisheries management as practised in Tonga in the 1980s focused on aspects of biology and economic development. This approach failed to prevent problems of declining fisheries resources. The management system has therefore been extended to consider social factors, in recognition that fisheries must convince resource users to conduct their activities in ways that are ecologically sustainable, economically viable and socially acceptable.

Traditionally, the people of Tonga have always had the freedom to harvest fisheries resources in any part of the sea without any restriction on species or quantity. In practice, judging from the evidence from Kolonga, subsistence fishers focus on the reef that is closest to home. Although there is no system of chiefly reef ownership as seen in many other Pacific cultures, some sort of unwritten agreement may in effect govern access to inshore reefs. This is an area that needs further research, as it would have an impact on the types of management strategies that would be deemed acceptable at the community level.

The modern approach to fisheries management highlights the importance of considering the social status of fishers when developing management strategies. In the case of the people of Kolonga, people are very much dependent on seafood for subsistence as well as income. They harvest anything that is edible or that has a cash value. For now, *Trochus* is collected only for food, not income. As long as a cash incentive to fish does not develop, there is hope that the *Trochus* harvesting ban can be made effective through an education program targeting fishers. If the attitude of Va'avao Mataele is indicative, an information campaign may suffice to deter many fishers from taking *Trochus*: "Like I said, I don't target *Trochus*, but when setting my net, if I see a *Trochus* I pick it and put it into the canoe…I never knew about the Ministry of Fisheries shellfish enhancement program….and I promise you I won't touch any *Trochus* from now on, because now, I know its not allowed to collect *Trochus*" (Va'avao Mataele, Kolonga village, age 58).

Gender and age are important concerns for *Trochus* management in Tonga. The majority of gleaners, who are the ones catching most of the *Trochus*, are female. Many of these fishers are either very young or very old. Any government consultation concerning the harvest of *Trochus* therefore needs to involve females, youth and the elderly. These fishers need to be consulted, not only because are they most likely to harvest *Trochus*, but because they are also among the most vulnerable members of society. Youth, women and the elderly often have few options for employment and depend heavily on fishing for subsistence. Their interests will be best served if the *Trochus* thrive and multiply. If they can be convinced of this, these fishers could be a powerful force for conservation and management.

The inclusion of fisherwomen in resource management will help ensure its success

RECOMMENDATIONS FOR ENSURING A FUTURE *TROCHUS* FISHERY

It is understood from the survey that the fishers who now collect *Trochus* have no idea about the *Trochus* transplantation program or the ban on harvesting. People of Kolonga need seafood, and their fishing ground has been over fished, which means that, left to their own devices, the fishers would keep on collecting *Trochus*. To achieve the goal of establishing a commercial *Trochus* fishery, the Ministry of Fisheries must strengthen its awareness program and persuade people to leave the *Trochus* in the sea. This will be a challenge because the most important breeders are the largest shells and these are also the most likely to be selected for eating. Therefore, focused communications must be directed at those who most fish *Trochus* for food.

In addition, some form of community based management needs to be implemented under which the community could take up the responsibility of protecting the *Trochus*. In other parts of the Pacific, periodic closures, protected areas, quotas and size restrictions are employed for managing *Trochus*. These measures often fail not because they are bad ideas but because fishers simply do not comply with the rules (Sims 1984, Tarisesei & Novaczek, this volume). To encourage fishers' cooperation, any local management system must be carefully designed with the participation of the primary stakeholders. In Tonga, these are the young people, women and elders who are most dependent on reef gleaning activities.

REFERENCES

Amos M. J. 1995. Combination of fisheries management regulation, traditional based management and wild stock enhancement using hatchery reared *Trochus* juveniles as a precautionary management principle for *Trochus niloticus* resources in Vanuatu. *SPC Inshore Fish. Mgmt. Bull.* 37.

Bell, A. J., 'U. Fa'anunu & T. Koloa 1994. *Fisheries Resources Profiles: Kingdom of Tonga.*

Fa'anunu, 'U. 2001. Recent surveys of transplanted green snail (*Turbo marmoratus*) and Trochus (*Trochus niloticus*) on Tongatapu, Tonga. *SPC Trochus Information Bulletin* #7, January 2001.

Foale, S 1998. Assessment and management of the Trochus fishery at West Nggela, Solomon Islands: an interdisciplinary approach. *Ocean & Coastal Management* 40:187–205.

Gillett, B.1993. The August 1992 transplantation of *Trochus* to Tonga and Niue. *SPC Trochus Information Bulletin* #2.

Gillet, R. 1995. Pacific Island *Trochus* introductions. *SPC Trochus Information Bulletin* #4.

ICECON 1997. Aspects of the industry, trade, and marketing of Pacific Island *Trochus*. *Pacific Islands Discussion Paper Series*, 2. Reykjavik, Iceland.

Manu, N., S. Shigeaki, & K. Udagawa 1994. Transplantation of commercial important gastropod mollusks, *Trochus* shell, *Trochus niloticus* and the green snail *Turbo marmoratus* to the Kingdom of Tonga. *ARDP Working Paper* No. 4.

Nash W. J. 1993. *Trochus*. In A. Wright & L. Hill (eds) *Nearshore Marine Resources of the South Pacific*. Suva: IPS, FFA, & ICOD. P.451–495.

Niumeitolu S, M. Mateaki & K. Kikutani 1999. Report on the seed releasing and re-stocking survey of *Trochus* and green snail. *Technical report of the JICA short term expert*. 11p.

Parkinson, B. J. 1984. *AReport on the Potential for the Introduction of Trochus to Tuvalu. Noumea.* New Caledonia: SPC.

Ponia, B. E. 2000. News from the Cook Islands. *SPC Trochus Information Bulletin* #6, January, 2000.

Sims, Neil 1984. *The Status of Trochus niloticus in the Cook Islands: 1984.* Unpublished report for Ministry of Marine Resources, Rarotonga, Cook Islands. 26p.

Zann, L. 1994. The status of coral reefs in the south western Pacific Islands. *Marine Pollution Bulletin* 29:52-61.

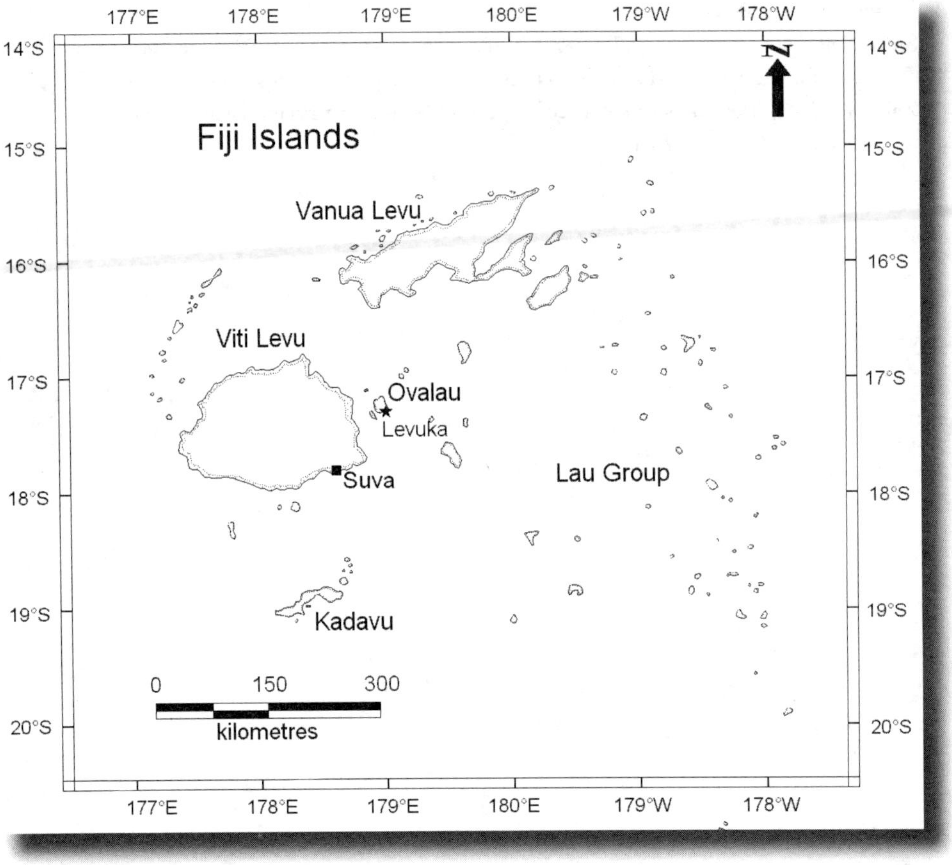

Map 1. Fiji, showing the location of Levuka on Ovalau Island

11 Gilt-edged Packet Or Economic Straight Jacket? A Case Study Of Cannery Workers In Levuka, Fiji Islands

Jyotishma Rajan

"Women are capable of dexterous work, better with their hands, more productive and more committed than the men."

— PAFCO Quality Control Manager, Levuka, 2000.

"PAFCO's annual wage increments are an insult to intelligence, to humanity...workers give the best years of their lives to the company but gain very little while the company takes so much."

— T. Tokalauvere, PAFCO workers union leader, 2002.

THE TUNA INDUSTRY IN LEVUKA

Levuka is a quaint town on the island of Ovalau, east of Suva, the capital of Fiji. Levuka was Fiji's capital until 1882 when Suva was declared the capital due to lack of space for expansion in Levuka. Levuka has now been declared Fiji's heritage town because of the many still intact relics from the colonial times. Levuka is rich in history; it is without doubt the birthplace of Fiji's modern economy and politics.[1] In fact, a journalist described Levuka as the place "where the first of everything began in Fiji" (Kalouniviti 2002: 6).

Levuka experienced major economic growth until the demise of the copra trade in the early 1950s. Young people started moving to Suva to look for jobs. Those who stayed back desperately sought alternatives to rebuild Levuka's economy and their livelihoods (Gibson 2002). Visionary citizens of Levuka, along with the then Japanese Minister for Industry and Trade were responsible for the subsequent establishment of PAFCO— the Pacific Fishing Company Limited (PAFCO 1999).

PAFCO was established in 1963 as a tuna trans-shipment base for Japanese tuna vessels in the Pacific. It began as a joint venture between the Fijian government and two Japanese partners, C. Itoh and Co. Ltd. and Nichiryo Ltd. The company employed men only at this time as stevedores, bait loaders and fish stackers. The government later realised the merits of adding value via canning and the trans-shipment base was subsequently upgraded into a cannery in 1970. The company still employed mostly men at this time (Anon, 1992: 35).

Later, women were employed to replace the men because the men refused to do monotonous and 'demeaning' tasks like cleaning, gutting and breaking the tuna into flakes to be canned (Emberson-Bain & Slatter 1995). Critics argue that, whereas managers pretend that women are simply better workers (see quote above), women are really the preferred group of employees because they are willing to work for low wages, follow orders easily, are more worried about being laid off and in many cases are not aware of their rights as workers (Lim 1980). Hiring women for the processing lines has worked, and continues to work perfectly for the tuna factory on Ovalau, where employment opportunities are severely limited. But even with this advantage, by 1987 the Japanese partners had withdrawn from the joint venture because it was not profitable. The Fiji government has been the majority shareholder ever since, with minority shares held by a few locals.

Since its establishment, PAFCO has been through boom and bust periods that are characteristic of many tuna ventures. Its successes were due mainly to successful integration into global markets through brand association (e.g. Cloverleaf, Bumble Bee, John West and Sainsbury's), preferential treatment under the Lome Convention and provision of raw materials and markets by foreign partners (PAFCO 1999). Failures were due mainly to poor management, periodic shortages of fish to process[2], industrial disputes, failures of IKA, a sister corporation and the impact of climatic variables such as El Nino (Rajan, MA thesis in prep.).

PAFCO employees also went through major boom and bust periods. For the workers, periods of 'boom' have meant secure jobs and acceptable standards of living while periods of 'bust' have meant intermittent lay-offs, factory closures and sudden decline in living standards. The most recent 'bust' period occurred between 1993 and 1996 when PAFCO almost closed down due to industrial disputes and the lack of fish (PAFCO, 1999). Relief arrived in 1997 when PAFCO signed a toll-pack agreement with Bumble Bee Seafoods, a US based tuna processing and marketing company (www.bumblebee.com, Mar 5, 2004).

Under the toll-pack agreement, Bumble Bee provides raw tuna for processing through Fong Chen Formosa, a global fishing company that provides all Bumble Bee plants with fish for processing. Bumble Bee also provides expertise and markets while PAFCO provides the facilities and labour for processing tuna into loins and canned tuna. Bumble Bee pays PAFCO toll-pack fees for processing the tuna.

With Bumble Bee arrived major changes. The factory was upgraded into a state-of-the-art tuna loining[3] facility and many benefits were put in place for the workers. Ever since the deal was signed, PAFCO has stepped out of unprofitability and financial insecurity and into high profitability and year-round operations (PAFCOs financial details could not be obtained). The agreement has provided employment for about 1000 people (about 12.5% of Ovalau's population), of which approximately 65% are women. PAFCO's good performance has also boosted business in Levuka town, which is almost entirely dependent on PAFCO and its workers.

But life for the workers has never been gilt-edged. Presented below are the stories of two women affected by PAFCO in two different ways. Their stories are testimony to the fact that capitalism does not necessarily result in holistic development, especially for those on the lower rungs of the 'class' ladder. Benefits that accrue to PAFCO from its involvment in the global tuna trade rarely 'flow down' to its workers.

CASE STUDY METHODS

Information for this case study was obtained in Levuka in March and April 2002 during data collection for my Master of Arts thesis. The main method of data collection was personal interviews using semi-structured questionnaires. Secondary data sources included newspapers, journal articles and other published and unpublished work. The two stories given below give a fairly representative reflection of the stories of the more than 30 women I interviewed at PAFCO.

LIFE AND WORK AT PAFCO: SALOTE'S[4] STORY

I had just received permission from the PAFCO Personnel Manager to speak to the workers. She told me to conduct my interviews in the canteen. "You cannot enter the production area", she had said. "The line supervisor, Miliana will send the women out to you". I had agreed. It was better to get the chance to speak to some workers than not speak to anyone at all. Plus, I could not interfere with the women's work schedule.

I made my way to the canteen and met the factory's nurse at the en-

Foreign-registered tuna ships come in to Suva harbour for fuel and supplies. Photo courtesy of Suva Harbour Foundation.

trance, which was near the sick bay. The sick bay is spacious and airy, with a bed, a bench, a table and a chair. I saw a stethoscope on the table and some supplies: Panadol, Band-Aids, Gentian Violet, cotton wool, gauze, basic medical equipment. I asked the nurse to call Miliana while I waited on the bench. Mili, as they call her, arrived in her white uniform, boots and blue head-cap. I found out later that supervisors wear blue, quality controllers red and other workers, white head-caps.

Mili had been told to expect me and had been instructed by the PAFCO Personnel Manager to release women from her production line to talk to me. She left me sitting in the canteen while she went inside the production area to send women out to be interviewed. "Not a very friendly smelling place", I thought, "I wonder how the women work all day in such an environment". Looking around the canteen I noted that it was quite large, with ceiling fans installed. The side facing the road had huge glass walls with small windows on top that let some air in. Beside the entrance to the canteen were the credit co-operative store and the workers' credit union. Both were closed. The other wall, adjoining the production area, had large notices on it in English and Fijian. These notices contained rules for the workplace, such as forbidding women to work if they had colds or cuts on their fingers.

Mili emerged from the production area with Salote, a 45-year-old Fijian woman. She was smiling at me. I stood up to greet her, asked her to sit down and after thanking Mili, I introduced myself and my purpose.

Salote has been working for PAFCO since 1982. She lives in Natokalau Village, about five miles away. The main mode of transport in Ovalau is by small and medium sized trucks; almost every village has its own. Every day Salote pays $2 return fare for herself and also for each of her children who attend school in Levuka. Since she works Saturdays and currently has one school age child, this adds up to $22 (US $12.76)[5] weekly in transport costs. Some women travel to work all the way from the other side of the island, she told me. That is a four hour round trip. "Then these women must have to wake up really early in the morning to get to work on time", I exclaimed. Salote nodded.

Salote earns $1.75/hr. The legal minimum wage for temporary and permanent workers in Fiji's manufacturing sector is $1.65/hr while the legal minimum wage for casual employees is $2.02/hr (Fiji Islands Trade and Investment Board, www.ftib.org.fj, Mar 5, 2004). PAFCO has permanent, temporary and casual employees. For 8.5 hours of work per day, she receives $14.87, or $89.25 per week in gross wages (US $51.72)[5]. Deducting FNPF[6] (Fiji National Provident Fund) and transport costs, Salote is left with $65 per week. Salote is repaying a loan she took from the PAFCO Workers Credit Union in the beginning of the year to buy her son's school supplies. She is also repaying a stereo bought on hire purchase. After that, there are the groceries to buy. The Credit Union and stereo payments are deducted directly from her pay every week. Though this ensures that the loan and the stereo will be paid for, it does not give Salote any flexibility over how much she wants to pay back at any one time. Sometimes she needs extra money to pay for certain things but does not have enough. At the end of each week, after meeting all the expenses, she is left with only $15 in hand. "It was

THE LOGIC OF CHEAP, GLOBALIZED LABOUR

Large companies such as Bumble Bee (termed transnational corporations or TNCs) relocate to developing countries mainly to minimise production costs. In most cases, labour intensive parts of the production process (in this case loining (Iversen 1987)) are relocated to developing countries with abundant cheap labour (Lieten 2001, Lim 1980). For example, Bumble Bee's two largest processing plants are located in Puerto Rico and Ecuador (www.bumblebee.com, Mar 5 2004). While Bumble Bee does not pay workers at PAFCO, it does play a major role in how much workers are paid. In fact, Bumble Bee was looking for a low cost producer in the Pacific close to its major competitor, Star Kist in American Samoa (Tabureguci 2002). Bumble Bee pays PAFCO a fixed amount of toll-pack fees per year. Fees are based on PAFCO's production costs. PAFCO cannot raise any production costs, least of all wages, because if it does so, Bumble Bee will not pay PAFCO any extra fees (R. Ward, PAFCO General Manager Operations, 2002). PAFCO has to work very hard to control its costs and wages are one of the most convenient expenditures to control.

harder when all my children were in school. Then I would have only enough money to buy tea and sugar", she added. To top this off, the workers' rate of pay per hour is increased by only 1% per year, which in Salote's case amounts to 1.75c per hour per year. All workers who were interviewed mentioned this wage increment rate and it was confirmed by the PAFCO Personnel Manager.

"Are you able to save any money?" I asked Salote. "Now, yes", she replied. "Before, when all my children were in school, I couldn't". "Does your husband work as well?" I inquired. "No, he just works on the plantation", came the reply. "Oh, what do you grow on the plantation?" I prodded. "*Dalo* [a root vegetable], cassava, *bele* [a green leafy vegetable], bananas and *yaqona* [kava]",[7] she replied. "Do you eat these at home or do you sell them in the market?" I prodded further. "Normally we sell the *dalo* and *yaqona* as they fetch good prices. If we have any spare crops, we sell that too", she beamed back. Salote and her family are lucky and also wise. Many women I interviewed later said their families had stopped practising subsistence farming and fishing after the women started working at PAFCO. This not only resulted in increased pressure on the women to feed their families, it also resulted in families not consuming fresh, healthy local foods but increasing their dependence on processed foods. PAFCO families who rely completely on food they buy from town usually subsist on tea, sugar, flour, biscuits, bread, margarine (butter is more expensive) and ironically, tinned tuna (Personal observations in town and communications with workers, 2002).

"Does PAFCO not provide transport to and from work for its workers?" I enquired of Salote. "Only to the night shift workers. PAFCO provides free transport home to night shift workers after they complete their shift. In order to come to work, night shift workers pay for their own transport, just like us". Later in the year, when I spoke to the General Secretary of the PAFCO workers union, I was informed that under a Master Agreement signed between PAFCO and its employees, PAFCO was required to provide free transport to night shift workers, both to and from the factory (Tokalauvere, workers union general secretary, pers comm 2002). This did not seem, however, to be the practice.

I decided it was now time to move on to another topic. "Salote, are you a permanent employee?" I asked. "Temporary", she said. I found this a bit confusing. Temporary after almost 20 years of service? "Yes", she replied. "Why?" I asked. Salote said that she had never been made permanent; she was told she did not work well enough. The PAFCO Personnel Manager later informed me that PAFCO conducted productivity appraisals every week under the performance based remuneration system (PBRS). Those who met the required threshold were promoted to, or retained as, permanent employees. Those who failed to meet the targets set were demoted, retained as temporary or in the worst case, terminated from employment. The PBRS ensures the highest possible levels of productivity from workers. In other words, the PBRS ensures greatest returns to PAFCO at least cost and lowest returns to employees, especially those that cannot meet the required thresholds for promotion. An interesting point to note here is that while the productivity of women is tested under the PBRS, there is no such mechanism in place to monitor the productivity of the men working at PAFCO. All employees interviewed as well as the PAFCO Personnel Manager attested to this. This was perhaps due to the fact that women were the ones working on the production line. Whatever the reason, the result is that, under the PBRS, the risk of job loss is far greater for women than for men.

Another benefit to PAFCO of keeping employees temporary is that it keeps total wage costs low. Temporary workers have to be paid for shorter periods of time than permanent workers. Temporary employees work monthly 'contracts' compared to permanent employees who work yearly 'contracts'. Every time an employee is demoted to temporary status, his/her wage level reverts to the basic wage regardless of how many years he/she has worked for PAFCO. There are almost equal numbers of temporary and permanent workers at PAFCO: 45% and 48% respectively (Daunakelo, pers comm 2002). The number of years of service these two groups of employees have given to PAFCO is very similar (Personal communications with workers, 2002).

"Are there any casual employees as well?" I asked Salote. She said there were. Casual employees were hired on weekly contracts as and when the factory required. At the time of interviews about 40 casuals were employed by PAFCO. I later got the chance to speak to some of them. They were remunerated at the same rate as the permanent and temporary workers. However this is much lower than the legal minimum wage for casual employees in the manufacturing sector in Fiji of $2.02/hr (www.ftib.org.fj, Mar 5, 2004). If any casual employee showed a very high level of productivity, she might be promoted to temporary status and retained as an employee. Others would be laid off once the factory requirements were met. Casual employees are hired on a 'round the island basis' i.e. one village after the other. The chiefs of Ovalau had requested this of the PAFCO management to ensure uniform development opportunities around the island. The same practice is used with employing stevedores (required only when a ship arrives with tuna for offloading) (Navuetaki, PAFCO Public Relations Officer, pers comm, 2002).

"Has PAFCO given you any training Salote?" I asked. "Yes", she replied, "last year some of us were trained to administer first aid. We also learnt about the OHS.[8] They also taught us how to clean the fish properly and how to make the loins and vacuum pack them. We were paid for the duration of the training and received certificates too", she smiled as she replied. I smiled back at her.

It is common practice in other processing/manufacturing factories to fragment the production process into many small tasks (Wheelwright 1980). The incidence of fragmentation is higher in the production area (where women work) compared to the other two sections: raw materials and logistics and warehouse. Fragmentation reduces the need for formal training. No special skills have to be developed, except in the cases of maintenance and quality control personnel. Savings in the form of low or no training costs and the low wages payable to unskilled or poorly skilled workers contribute significantly towards PAFCO's cost reductions.

I had read earlier that work conditions had improved significantly after Bumble Bee Seafoods arrived. The canteen and sick bay were testimony to this. I had also read about the laundry, Credit Union, the Day Care Centerethat runs on the UNIFEM concept and the gym (Pareti 2001). I proceeded to ask Salote about each of these.

"Do you get your uniform washed at the laundry?" I asked. Yes", she replied, "The laundry washes all our uniforms for free and dries them too. We do not have to worry about the weather now as we had to before, when it rained. But there are some women who do not give their uniforms to the laundry because their uniforms got stained in previous washes. They take their uniforms home". A majority of the women get their uniforms washed at the laundry. The uniforms workers wear are now free; before Bumble Bee arrived, they had to purchase their uniforms from PAFCO and boots in town. The laundry was introduced for the purpose of maintaining hygiene, because in the past some workers would come to work in unwashed uniforms. As such, the laundry is beneficial to all the workers, PAFCO and Bumble Bee.

After Bumble Bee arrived, PAFCO also put in place a gym where the workers could exercise. "What about the gym Salote, do you use it?" was my second question. "No", she replied dejectedly. "Why not? Are you shy?" was my next question. "No, it's too far. Lunch is only one hour and by the time we change into gym clothes here, walk to the gym, work out in the gym, come back to work, change back into our uniforms, it will be past lunchtime. Plus most of us are tired by lunch and we just want to sit down and relax," said Salote. "What about using the gym in the afternoon?" I was inquisitive.

"Oh no, no. In the afternoon we have to run to our carriers after doing our shopping in town. If we go to the gym then we will miss our transport" came the reply.

I wondered if this was the case with all the workers. Further investigation revealed the following. Some of the ladies felt that they were too old to do heavy exercise in the gym. They preferred walking every day to exercise their tired legs. A greater proportion of men used the gym; they are less likely to have to return home to attend to household chores. Men are also paid more than the women and so can afford the $0.50/day gym fee. Interestingly, none of the men who do not use the gym cited high fees as the reason; women on the other hand, said the fees were too high.

"What about the doctor from Suva, Salote? Do you get yourself examined by him when he comes?" I asked. The Personnel Manager had told me that PAFCO hires a medical doctor from Suva once a year to examine all workers. She had also said that many women were too shy and did not see the doctor when he arrived. I had my suspicions about this, which were confirmed by Salote's response to my question. "Yes, I see the doctor every time he comes to PAFCO. But some of the women do not do that. They are frightened. They think that if the doctor feels they are too sick, he will tell them not to come to work and they will lose their jobs."

"What happens when workers are found to be very sick?" I wanted to know. "Most of the time they are told to go home", she replied. "Do they receive any compensation if their sickness is because of the work they did?" I asked further. "No", she responded, "We know that we should receive compensation for any sickness or injuries at work but the company never pays us. So many women now have bad backs and knees because of standing up for long periods. Most of them are home now with no money for treatment". "What about pregnant women?" I wanted to know about maternity leave conditions. Salote shook her head as she replied, "Most of the women hide their pregnancies for as long as they can. As soon as the management finds out they are pregnant they are sent home. They are not paid the maternity leave allowance unless they are friends with the management".

I later met a woman who had been sent on unpaid maternity leave. She had been a permanent employee before she left. When she returned after giving birth, she was told to start as a temporary employee on a lower wage rate than when she left. When she asked why she should have to do so she was told it was company policy. PAFCO also has a practice of telling workers (men and women) to take shorter than recommended sick leaves. As a result workers return to work sick, which results in productivity losses. These workers then face the risk of demotion, or in the worst cases, termination for poor performance (Personal communications with workers, 2002). The PAFCO Workers' Union General Secretary backed up this claim and while the Personnel Manager refused to elaborate, she did state that management did not encourage workers to take longer than necessary sick leave.

At that moment the door linking us to the production area swung open and out stepped about 60 men and women for their morning tea break.

Some bought tea and bread from the canteen. All sat down to rest. "You should get yourself some tea too Salote," I suggested. So she went to get her tea while I looked over the notes I had made so far. While Salote was away I saw an old lady walk across the canteen with a horrible limp. Later I asked the reason for her ailment. "Oh her, she is Paulini. She was working for PAFCO long before I joined. She used to work in the production lines before. One day her knees became so painful that she could not stand up for even an hour," Salote replied. "Oh dear!" I exclaimed, "What happened then?"

"She was a good worker and maybe the company did not want to lose her so she was redeployed to the warehouse. Now she does piece work," replied Salote. I was later informed by some of the workers as well as the Personnel Manager that piece workers are paid for each item they put together. Paulini receives \$0.07 for each cardboard carton she glues together. When she was working in the production area she probably earned around \$1.75/hr. Now, in order to earn the same \$1.75, she must glue together 25 cartons in one hour. She achieves this feat with much difficulty. Although she does not have to stand up, she still needs strength to ensure that the flaps of the carton have been glued firmly.

Paulini is similar to other piece workers who are paid one cent for every can they label. Ironically, labelling at PAFCO is mechanised, which means that only those cans that have been incorrectly labelled by the machine can be re-labelled by the workers. In order to earn \$1.75, these workers must label 175 cans in one hour, an opportunity they rarely get. Piece workers do not receive any FNPF deductions, while other workers do (Tokalauvere 2002). I could not help thinking how unfair work and life was for some PAFCO women.

Tea break was over; Salote had to go back to work. As I thanked the kind Salote and bid her goodbye, I noticed that I smelled like a can of tuna but felt I had no right to complain because of the many mothers, daughters and grandmothers who perpetually smell of tuna because they have no other choice.

LIFE AFTER PAFCO : MERE'S STORY

I met Mere in her house in Draiba village, about a 500m walk south from PAFCO. Mere is married and is a young mother of four. She was sitting inside the house with her husband. A woman from the village who had agreed to help me introduced me to the couple. They invited us inside. As I spoke to Mere and her husband, I looked around their small wooden house. We were in the middle of the house with a small kitchen on one side and the sleeping area on the other. Some schoolbooks were on the floor and some clean clothes on one of the beds. Though small it was clean and neatly kept. The contents of the house gave me the impression that the owners kept only as much as they needed to get by, no frills or fancies.

Mere used to work for PAFCO but she was fired from work in 1997, after participating in a strike. She has not been employed since. Mere is one of 57 employees terminated in 1997 for what management termed lax attitude and poor performance (Nadore 1997). The workers fought and won their case in court. In January 2002, after the Arbitration Tribunal ruled in the workers' favour, the High Court ordered that they be reinstated with full pay. It ordered PAFCO to re-employ 24 of the 57 workers immediately and pay them wages outstanding. The remaining 33 were to be reinstated when work became available. At the time of research, the remaining 33 had not been re-employed. Mere was one of them.

Mere's income from PAFCO used to supplement what her husband could earn selling *yaqona*. Their standard of living was much higher when she was working at PAFCO. She used to earn around $60 every week which, when combined with her husband's income, was just enough to meet their family's needs. Luckily for Mere and her family, her husband owns a farm on which he grows cassava, *dalo, bele* and *rourou*[9] for subsistence and *yaqona* for sale in the market. Mere and her family somehow manage to survive on whatever they earn from selling produce in the Levuka municipal market every Saturday.

A few years ago they used to earn a lot of money from selling *yaqona*. This income declined drastically after European countries began to complain of liver diseases attributed to consumption of kava tablets. With more advanced research pointing out that kava tablets can be safe, the fortunes of families like Mere's may improve.

"How many children do you have?" I asked Mere and her husband. "Four," they replied. "All of them are in primary school." The primary school their children attend is not too far from Draiba village. As such they walk to school so Mere and her husband do not have to worry about the cost of transportation. They do, however, have to worry about four school uniforms for rapidly growing children, four sets of school stationery and other requirements and four lunches every day. "What do your children take for lunch?" I asked. "Sometimes when my husband or I go fishing I give them cassava and fried fish. My children, they do not like taking boiled fish to school as the soup tends to drip from their lunch boxes. Other times I give them cassava and rourou, or tinned fish and rice. And sometimes, only sometimes, I have to give them noodles and tinned mutton for lunch when we do not have anything else to eat at home," she replied.

"Have you bought anything on hire purchase?" I asked. Mere looked crestfallen as she answered, "I had bought a stereo and a gas stove when I was working at PAFCO. But they were repossessed after I could not keep up with payments after being laid-off. Now I use a kerosene stove." "Does your village have electricity?" I asked. "Yes it does but not all of us have electricity in our homes because we cannot afford the bills. We use a benzene lamp at night," Mere's husband replied. I realised that Mere and others like her had accepted that their dreams would probably never come true. Mere's eldest will start attending high school in three years' time and she and her husband have a tough battle ahead of them to provide for their children.

At that moment my eyes caught a beautiful crochet work on the sole armchair in that room. "Wow, did you make that Mere?" I asked. "Yes," replied her husband proudly. "Sometimes I tell her to make more things like that and try and sell them." She looked at her husband with a shy smile saying, "Yes, I could but not many people want to buy such things all the time. Maybe I should try some other craft as well". We all smiled. Levuka is a very small town which boasts two supermarkets, two restaurants, a shoe shop and some mixed trading shops. The market for luxury items and handicrafts is very small, so relying on handicrafts for some income would be precarious. The lack of economic diversification in Levuka pushes people to rely on PAFCO for employment. They are then very vulnerable to the consequences of being laid off.

I had come to the end of my interview. I wished Mere and her husband luck, thanked them warmly and left.

THE NEED FOR CHANGE

Workers such as Salote and Mere depend on PAFCO for their livelihoods. PAFCO in turn depends on Bumble Bee Seafoods for its existence. By depending on Bumble Bee, PAFCO and its workers, especially women, are vulnerable to changes in the competitive global tuna trade. This is typical also of other manufacturing and processing industries, especially in developing countries. The stories of Salote and Mere are not unique; they reflect the lives of their counterparts in PAFCO and elsewhere in the world.

By depending on Bumble Bee, PAFCO and its workers also become vulnerable to the company's profit-driven practices. This article has highlighted clearly how Bumble Bee influences the level of wages paid by PAFCO. Fear of losing its valuable but dominant partner keeps PAFCO from improving wages and conditions for its workers. To become and remain competitive in the global tuna trade, the highest level of productivity and cost effectiveness are required. The former is gained through the PBRS, PAFCO's reluctance to provide sick and maternity leave and 24-hour operations. The latter is attained through low wages, scant training opportunities, stringent promotion rules, a preference for temporary employees. Miserly wage increments and fragmentation of production.

Sadly, PAFCO's bid to remain competitive and Bumble Bee's attempt to remain profitable and have a presence in the Pacific have not resulted in anything constructive or beneficial for the workers, especially women. They are still poor, they still suffer from arthritis of their knees, they are still discriminated against for being female, and they are not provided the work benefits they deserve. The fact that none of them pay taxes shows that all live below the poverty line. The requirements of a global industry have profound impacts on women in small communities in developing countries.

PAFCO women need to be paid higher wages for all the hard work they do. PAFCO should also uphold provisions in the Employment Act and provide women paid sick and maternity leave. The company needs to accept its responsibilities under the Workmen's Compensation Act, and respect the Master Agreement it signed with the workers, as well as Court decisions. The government should not compromise the welfare of Fijian workers in the name of global competitiveness. The government normally defends itself by saying that PAFCO in Levuka is performing a social service by providing employment to many, especially women. If the government and industry were really interested in doing a social service, then they would look deeper into the concerns of PAFCO women. The current government could, through its affirmative action policy for indigenous people, at least improve the welfare of indigenous PAFCO workers, who make up 96% of the labour force at PAFCO (Personal communications with workers, 2002).

There is also a need for economic diversification in Ovalau. This has been an ongoing task and should be given priority. Another large employer on Ovalau would provide alternative employment and would also influence the wages paid at PAFCO. PAFCO's near monopoly position (the largest 'buyer' of labour in Ovalau) allows it to set its own wage rate, which does not reflect the cost of living in Ovalau. This situation needs to be rectified. There is also a need to assess government development policies, especially where women are concerned. Developing countries should no longer be considered, nor behave, as the low-cost manufacturing centres of the world.

To conclude, if the social cost of growth-driven and export-oriented development are not addressed, women such as Salote and Mere will continue to suffer unnecessarily.

NOTES

1. 'Modern economy and politics' here refer to the economics and politics of the colonial era, as opposed to the traditional, largely subsistence economy and chiefly political system that existed in pre-colonial times.
2. Lack of fish was a result of a combination of factors. At times PAFCO did not have enough money to pay its suppliers. Other times it failed to meet the derogation requirements of the EU market. Under the Lomé Convention, ACP countries must ensure that at least 90% of the products they export to the EU is local content. To meet this requirement, the government had established IKA Corporation (later made a subsidiary of PAFCO), a long-lining tuna fishing company responsible for supplying raw tuna to PAFCO. IKA failed due to poor management, small number of vessels and outdated technology.
3. Loining refers to the process of creating tuna loins. The tuna is skinned, de-boned and divided laterally into four parts (called loins).
4. Real names of PAFCO employees have been withheld to protect their anonymity, as promised.
5. At the time of writing (2003) FJ$ 1 was worth US $0.58.
6. Fiji National Provident Fund, a form of insurance that assists workers who lose their employment and pay benefits to families when they die.
7. *Yaqona* refers to the root of the *Piper methysticum* plant that is dried, ground into powder, mixed with water and consumed as a national drink in Fiji. *Yaqona*, also known as *kava*, is a highly priced commodity therefore a lucrative source of income.
8. Occupational Health and Safety Act.
9. *Rourou* refers to the leaves of the taro (*dalo* in Fijian) plant. *Rourou* is prepared by boiling leaves into a fine paste and is one of the staple Fijian foods.

REFERENCES

Anonymous. 1992. PAFCO- and its origins. *Pacific Islands Monthly* Dec.: 35.

Emberson-Bain, 'A. and Slatter, C. 1995. *Labouring under the law: a critique of employment legislation affecting women in Fiji.* Fiji Women's Rights Movement, Suva.

Iversen, R. T. B. 1987. *US Tuna Processors.* Multinational Corporations in Pacific Tuna Industry Project. Pacific Islands Development Program, East-West Centre, Honolulu.

Kalouniviti, J. 2002. Stepping Back in Time in Levuka. *Fiji Times* Apr. 13: 6.

Lambeth, L. and Leniston, M. c. 2000. Fiji Tuna Development and Management Plan. Vol. III Supporting Documents. Chapter Four: Gender Issues and Social Impacts. Jointly published by Secretariat of the Pacific Community, Pacific Islands Forum Secretariat and Canada South Pacific Ocean Development (C-SPOD) Programe, Phase II. Draft.

Lieten, K. 2001. Multinationals and Development: Revisiting the Debate. In Schurman, F. (ed.), *Globalisation and Development Studies: Challenge for the Twenty-first Century.* Sage Publications. pp. 99-115.

Lim, L. Y. C. 1980. Women Workers in Multinational Companies: the Case of Electronics Industry in Malaysia and Singapore. In Kumar, K. (ed.), *Multinational Enterprises: Their Impact on Third World Societies and Cultures.* Westview Press, Boulder.

Nadore, I. 1997. 84 cannery workers sent home. *Fiji Times* Jan. 8: 5.

Pacific Fishing Company Limited 1999. Briefing for the Right Honourable Prime Minister Mr. Mahendra Chaudhry at the Prime Minister's Office in Suva at 5pm on Wednesday the 6th of October 1999. Unpublished PAFCO document.

Pareti, S. 2001. In the name of the truth. Sun Weekender. *Fiji Sun* Nov. 24: 8-9

Rajan. J. in prep. MA Thesis, University of the South Pacific.

Tabureguci, D. 2002. PAFCO rides on brand association. *Daily Post* May 22: 8.

Wheelwright, E. L. 1980. *Transnational Corporations and the New International Division of Labour. Some Implications for Australia. Transnational Corporations Research Project Occasional Paper No. 4.* University of Sydney, Sydney.

www.bumblebee.com. 5 Mar 2004.

www.ftib.org.fj. 5 Mar 2004.

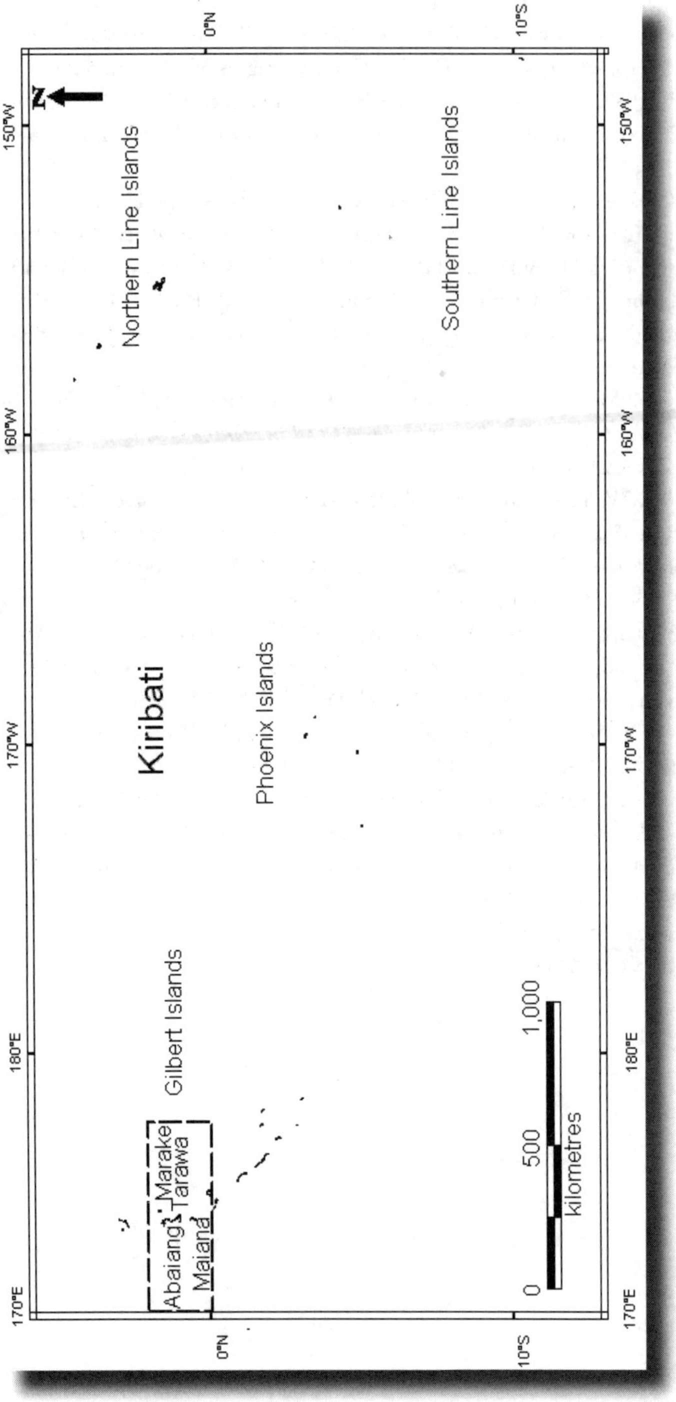

Map 1. Kiribati, showing location of the Tarawa Atoll.

12 Selling Fish In Tarawa, Kiribati

Maere Tekanene

"I am very happy because I have some money every day. I wouldn't know how I would meet ends without this job."

— Female fish seller, Betio

"We do not know anything about the Voluntary Kiribati Provident Fund nor the employer's duty to accommodate that. No one has ever told us about it."

— Female fish seller, Betio

WOMEN'S CONTRIBUTIONS TO FISHERIES IN KIRIBATI

Women occupy important but scarcely recognized positions in the inshore fisheries of Kiribati. They gather a variety of inshore resources for family consumption and are key labourers in the inshore commercial shell fisheries (Fay-Sauni & Sauni, this volume). Women predominate in the labour force that sells fin fish in domestic markets (Mitchell 1994, Tekinatiti & Wichman 1995). In this paper I shall explore women's work in the fish trade on Tarawa, the most populous of Kiribati atolls. As the first quotation above suggests, some women fish sellers may turn to the fish trade out of desperation, having few or no apparent alternatives to generate income. Women who do such work in the informal sector face many problems for they are often vulnerable to exploitation and may be neglected by employers and public officials who should ensure their welfare.

In Kiribati, fishing is of central social, economic and cultural importance and while the sale of fish is now commonplace, it is a relatively recent development and therefore reflects changing socio-economic conditions. Women's work in fish trading is crucial to distributing the catch in Tarawa, a densely populated atoll where ever-smaller numbers of the population fish for a living.

I shall briefly describe the range of women's work in fisheries and focus on the problems faced by women fishmongers. This will provide an understanding of one part of women's work in the fish trade that is, I suggest, vital for maintaining food security and generating income on Tarawa.

CASE STUDY METHODS AND OBJECTIVES

This study is a focused inquiry into working conditions and other factors that influence the fish trade on Tarawa, Kiribati. The study is action research for it seeks to identify opportunities for advocacy to improve the working conditions of women involved in the fish trade. Town officials were interviewed both to gather information about regulations and policies and to begin the process of encouraging them to pay attention to conditions of work in the fish trade. Fish traders were interviewed with assistance from two members of the Kiribati National Council of Women (KNCW). The purpose of involving the KNCW was to expose the council members to the issues in the informal fish trade, where the majority of employees are women. The interview process also sought to raise awareness among female fish traders concerning what they can do for themselves to improve their work conditions.

Data were gathered in August and September 2003 through a series of structured interviews conducted with 15 small-scale fish sellers, three town council officials, one fisheries officer, one Chief Councillor, an officer from the Kiribati Provident Fund and one owner of a small fisheries business. The areas covered were South Tarawa's main towns of Betio, Bairiki and Bikenibeu.

The KNCW volunteers met to finalize the interview questions and were assigned geographic areas and a target number of women to interview. Interviews were designed to document the socio-economic position of women engaged in the fish trade, their conditions of work, knowledge of workers' rights and employers' responsibilities and the social and environmental factors that influence their ability to gain a livelihood.

SOUTH TARAWA, AN URBAN ATOLL

Kiribati, which has been independent since 1979, comprises 33 low-lying atolls that are divided into three groups: the Gilberts, the Line Islands and the Phoenix Islands. The total land area of Kiribati, 822.8km^2, is spread over 13 million km^2 of ocean. It supports a population that rose from 72,335 in 1990 to 84,494 in 2000. Women have outnumbered men for the last three censuses.

South Tarawa, the capital of Kiribati, is located in the Central Gilberts Group and occupies a land area of 15.76 km^2 (Kiribati census 2000). It is home to 36,717 people, or 43% of the national population. Households are large, with an average of 8.1 people living in each home (Kiribati census 2000). The main three towns in South Tarawa, where government offices were set up during colonial British rule, are Betio, Bairiki and Bikenibeu. Betio is the most densely populated area, followed by Bikenibeu (Table 1). The dense population of South Tarawa provides opportunities for business development while improvements in transportation have given people easy access to the businesses. Being the main centre for shopping and work, Bairiki is most heavily visited during daytime. Bikenibeu also has major shopping centers and offices that are visited by people from other villages.

Increasing population on Tarawa, including in-migration from outer atolls, has filled any free space between villages in the atoll. In the urban sprawl from Betio to Tanaea there are now 11 main villages. With urbanization, land has become increasingly monetized and privatized. Many people indigenous to South Tarawa have sold their land to outer islanders who have come in search of cash employment, better education and access to health services.

The rapidly increasing urbanization of South Tarawa is linked to the increasing importance of and reliance on the cash economy. There appears to be a growing preference for an urban life among young people and among adults unable to make enough money from their traditional activities on outlying islands. This represents a significant shift in attitudes to paid employment. In the 1960s and 1970s people were not prepared to work for a private business. A common phrase normally used when speaking to an employed person in those days would be 'why waste your energy helping to build someone else's property?' This connotes a degree of jealousy but at the same time it conveys the self-reliant outlook of people at a time when money was not the only means of ensuring survival. Today, as the need for money increases, the reluctance to work for a private business or another family has diminished. Along with this changed attitude is a willingness to work for money alone, without demanding other financial benefits and work entitlements such as are enjoyed by members of the public service.

Table 1. Population of main towns and other villages on South Tarawa in the year 2000.

Village / Town in South Tarawa	Population Females	Population Males	Total	Total Households	Average Household Size
Betio	6218	6050	12268		
Bairiki	1386	1278	2664		
Bikenibeu	2945	2668	5613		
Other villages	7273	8899	16172		
Total	**17822**	**18895**	**36717**	**4529**	**8.1**

GENDER ROLES IN SMALL SCALE FISHERIES

Men's Reef and Deep Sea Fishing

Whereas women in Kiribati are very active fishers of inshore areas, men typically go fishing in the deep sea or along the outer edge of reefs. In South Kiribati, when a man comes home from deep sea fishing it is customary for his wife and children to meet him on the beach to help take home his fishing gear and catch. Tending to the catch and distributing shares to the nearest kinfolk (normally living as neighbours) is a woman's role. If a husband has borrowed a canoe or fishing gear, the wife ensures that the canoe owner is compensated with a share of fish that is equal to that taken by the fisherman. This is known as the canoe share or *tibwan te wa*.

Men use a variety of fishing technologies, including deep water hand lines, fishing rods, gillnets and specialized pole and line gear used for tuna. When men go fishing along the reefs, pelagic species such as flying fish are the most sought after. Spearfishing and octopus fishing are also commonly done on the reef.

Small-scale fishing businesses targeting small tunas employ men as crew. These businesses also operate on the traditional share system but instead of getting a share of the catch, the fishermen each get a share of the cash from the sale of fish. The crewmen take their shares only after all boat expenses have been paid, e.g. petrol, cigarettes and food. If there are three fishermen, the profits are divided by five: one share for each fisherman, one share for the engine owner and one for the boat owner (usually these are the same person). Sometimes catches are so low that the returns are not enough to cover expenses and then these expenses are subtracted from the next day's fishing. The money earned per week may be as little as$60, less than what women can earn from selling fish. Women who sell fish for such a company get paid according to the weight of fish sold. High sales are rewarded with bonuses.

Women's Inshore Fishing Activities

Women's inshore fishing is most frequently done in lagoons and includes a wide range of fishing gears and activities (Table 2). Reef and lagoon gleaning activities include the collection of a variety of shellfish, invertebrates, small eels, octopus and small fish trapped in tidal pools or lurking under rocks on the reef flat or lagoons (Luomala 1984: 554). Gleaning shellfish is women's major fishing activity because it can be done close to home, takes relatively little time, requires no costly fishing equipment and may be done in the company of children. Women's traditional knowledge helps them to identify and locate the shellfish on the seabed so they can collect plenty in a short time.

In recent years the most common target shellfish for women in South Tarawa is *te nouo* (*Strombus luhuamus*). This fishery has become more common as the *te bun* (*Anadara holoserica*) stocks that were predominant targets in the past appear to be depleting (Fay-Sauni & Sauni, this volume). Other shellfish harvested near villages such as Bonriki and Ambo in South Tarawa are *te koikoi* (*Asaphis violascens*) and *te koumwara (Gafrarium pectinatum)*. No fishing equipment is needed apart from a sack or a basin to carry the catch. Many children accompany their mothers in this activity, as it is easy, safe and fun.

Reefs, although important fishing areas (Table 2), are less commonly gleaned by women compared to the lagoon mudflat and shallow areas, because reefs are distant from villages. Lagoons are closer and appear to be more productive. Also, reef gleaning may require boots to allow walking on rough and sharp coral.

Table 2. Fishing habitats, types of fishing activity and women's participation in South Tarawa fisheries and aquaculture.

Habitat	Type of activity	% Fishers who are female
Lagoon	Shellfish gleaning	95 %
	Seaweed farming	at least 80% in outer islands
	Torch fishing	75 %
	Small rod fishing	35-50 %
	Handlining	5-10 %
	Gillnetting	5 %
Reef	Hand collection	50 %
	Rod fishing	5-10 %
	Gillnetting	5 %
	Lantern fishing	3-5 %
Ocean	Pole & line fishing	0 %
	Trolling	0.5 %
	Drop lining	0.5 %

Source: Tekinatiti & Wichman 1995.

Women also engage in a variety of other fishing activities, which are briefly noted here. It is primarily women who spear *te ibo*, a mud worm (Sipunculid). This activity is called *te waibo* (Luomala 1980:538). Both men and women go torch fishing (*te kibe)* at night. On a calm day or night they go rod fishing (*te roaroa*) (Catala 1957, Mitchell 1994). The fish commonly caught are *te ninimai* (silverbiddy) and small *te rereba* (trevally).

Te karaun, shallow water seining, may involve more than two people. The net is spread out in a semi-circle and the participants frighten fish into the net by running towards it while throwing stones and hitting the surface of the sea. *Te karaun* can be done in the lagoon or on the reef but it is mostly used in the lagoon where there are no rock pinnacles and coral to entangle the net. The owner of the fishing net gets a share of anything caught in the net; this share is called *tibwan te karaun*. A fishing net is not easy to buy because the cost is high. The Fisheries Department has in the past made special orders that are more affordable, so as to help families with low income to acquire a fishing net.

Recently, as people have become more desperate to catch fish, the reef side in South Tarawa has become more intensively used, especially during calm weather.

DEVELOPMENT OF THE TARAWA FISH TRADE

Origins

Prior to about 1960, inshore fishing on Tarawa was exclusively for home consumption and for meeting family obligations; fish were not publicly sold. Since the 1960s women have become involved in the fish trade, selling fish from home and walking around the village with their handcarts of fish. In the past, families and fishing groups would signal the availability of fish for sale to the neighbourhood by ringing a bell.

The Betio Fish Market

In the mid 1970s the first fish market was set up in the central part of Betio to cater for fish traders. Funds for building the market were provided by the Government of Kiribati, which gave the Betio Town Council the responsibility for operation and maintenance. The building had proper sanitation when it was first set up. Water was available for washing hands and drainage was installed in an open area in front of shops that shared the same building. Tables made of flat concrete slabs raised on layers of concrete bricks were used to display the fish because at that time few people owned ice boxes (locally called eskies). Fresh fish were spread on the tables and women would sit on stools also made of concrete, encircling their tables

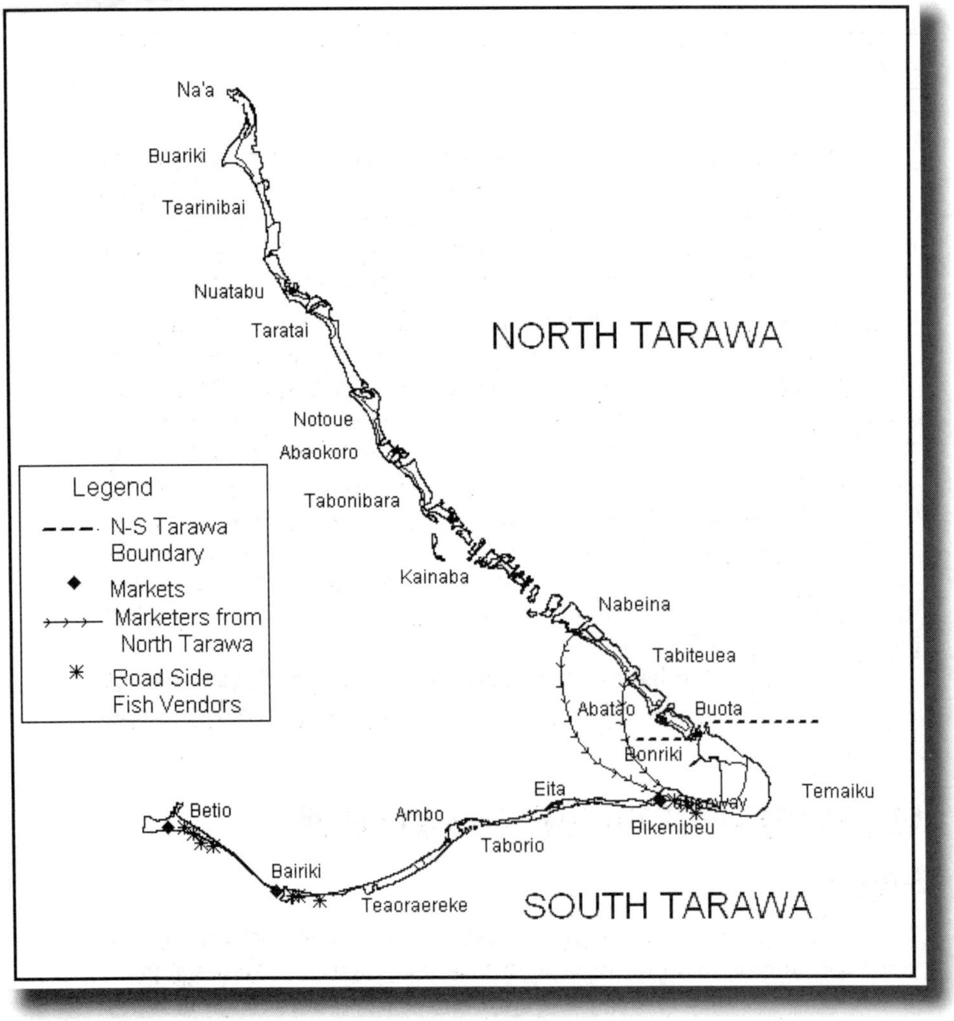

Map 2. South Tarawa, showing the location of various fish markets.

and waiting for customers. At that time, fresh fish was in high demand and generally sold out. In the early 1990s, the Betio Town Council allowed some new businesses to close part of the fish market and use it as a garment factory and retail shop. This reduced the once spacious area and forced women to crowd into one side of the market facing the main road.

The Bairiki Market

In the late 1980s, government built the Bairiki market. Located at the old ferry station at the end of Bairiki facing Betio, this location used to have the advantage of crowds of people waiting for the ferry. Since the construction of a causeway to Betio, the area is now much less popular and thus is no longer a good site for marketing fish. Bairiki, unlike the other two towns, has a small resident population so most potential customers are present in the daytime when they come for work, shopping and other errands. The key problem with the Bairiki market is the distance from the busy shopping centre and other highly visited places.

The Bairiki market building is enclosed and its interior was properly furnished with tables that drained water to one side. When this building was completed, the Tarawa Urban Council (TUC) announced to all fishmongers that their market was ready and that all those selling fish were required to use it. Despite various reminders, not many women took interest in the new facility. They stubbornly remained on the roadsides, believing that this was more convenient to their customers. The TUC failed to pass any official bylaw to enforce their plan, and they have never carried out a community consultation to encourage support for the venture.

Bairiki fish market in front of the National Library. Fishing families have constructed their own shelter by the side of the main road.

The Bikenibeu Market

In the early 1990s the market in Bikenibeu was completed with funding assistance from Canada. The market, located in the main business area of Bikenibeu opposite a big shop and the bank, has an outside toilet, an office for council personnel, and tables and seats made of concrete. However, there has never been any drainage even though it was built as a fish market. The design did not accommodate fish traders' basic needs. A concrete water tank installed on the south end of the building has not been in use for a long time because it developed a leak and was not repaired. Women selling fish here must bring water with them. On some occasions, when women fail to bring buckets of water, they repeatedly dip their hands in the bloodstained esky water. They serve customers without ever washing their hands properly.

The back of the Tarawa Urban Council fish market at Bikenibeu, showing how fencing has been used to enclose the area.

On the front of the Bikenibeu market building there is a signpost saying TUC Fish Market, but selling fresh fish is not permitted inside the building. Women are allowed indoors only to sell dried fish, shellfish and cooked products. In 2002 conditions at the Bikenibeu market were made even worse when wire fencing was installed to prevent night trespassing. This has impeded customers' access to the market.

CURRENT TRADE IN SHELLS AND FISH

Selling Shellfish in Tarawa

In addition to the finfish trade, some women engage in selling shellfish, especially the ark shell *te bun* (*Anadara*) and *te nouo* (*Strombus*) that they harvest themselves. Before overcrowding became a problem in South Tarawa, families penned *Anadara* on the adjacent mudflat, encircling them with big stones. They could use them as needed. Nowadays, people prefer to sell their shellfish immediately in the markets and the practice of sea farming or penning is a thing of the past.

Cooked shellfish destined for the market are placed in a plastic container with some pieces of coconut or cooked breadfruit and fermented toddy is added as vinegar. These are sold mainly inside the market building at Bikenibeu. Some are also sold in an open-air market in the Bairiki shopping area.

Since an outbreak of cholera in the late 1970s, few *Anadara* from South Tarawa are sold in bags; instead, presumably uncontaminated shellfish are brought from North Tarawa. The sellers of shellfish are therefore women from North Tarawa. They travel south in village canoes that cost AU$2.40 per round trip. At the market, vendors are charged AU$0.50 per market day. Their daily earnings usually run between AU$10 and $20. Regular marketers organise what they call *te karekare* or credit schemes through which women can get enough money to purchase a piece of equipment that they otherwise could not afford with their small earnings. Every day, one participant is given AU$5 and on the next day another is given the same amount. The rotations continue, with some deciding to pull out at the end of the first round.

In recent years, concern has increased regarding sewage contamination of the shellfish that grows in the Tarawa lagoon. This has affected the market acceptability of shellfish and has deprived families of this protein source as well as the money that could be earned by those who lack other employment. More recently some fishing areas in Tarawa lagoon have been reportedly affected by activities of locally owned sea cucumber companies based in South Tarawa. In Buariki, the villagers fear that sea cucumber gutting activities in their lagoon have in some way affected the habitat for *Andara* and caused them to disappear. Despite their concern, women have never lodged any complaint about this activity or about the disappearance of shellfish that were an important source of dietary protein.

In some islands of Kiribati, the municipal councils set regulations that restrict visiting boats from other islands from undertaking any fishing that would deplete the local fish resources. The North Tarawa Council has not raised this as a concern, nor has the women's NGO that is affiliated with the council.

A Profile of Tarawa's Fishmongers

Women sell most of the finfish sold on Tarawa. In total there are perhaps 60 fishmongers active in the municipality of Tarawa during periods when fish catches are high. This research is based on data drawn from approximately one quarter of this population. The women came mainly from families that lacked formal employment. Their ages ranged from 25 to 60 years old. On average, they were caring for households that contained 10 members, including three children.

Of the 15 women interviewed, one woman had no formal schooling, half had only primary education, one had lower secondary schooling and the remaining third had reached upper secondary school. Most of the women were working to supplement the cash income earned by a spouse. It was common for a woman to sell fish for the same small company that employed her husband as a fisherman. In other cases the spouse had only a temporary job, or there was no other breadwinner in the household. More than 80% of the respondents claimed that much of their income was spent on food and other basic family needs.

More than half of the women that were interviewed had worked in the small scale fisheries business for more than a year; it was their only means of earning cash. All said they enjoyed what they were doing mainly because of

A DAY IN THE LIFE OF A FISHMONGER

Mrs Teauama has been selling fish for the same fishing company for 11 years. She is in her early thirties, is married and has three children. The eldest child is staying with relatives in another village and the two living with her are aged one and 11. She and her family live in Bikenibeu where her husband grows cabbages around their house to supplement her earnings. At the time I visited her at 8pm on a Monday evening, she was just lighting her open fire to cook the evening meal. Her husband was out joining some Bahai lessons in Bikenibeu town. The baby slept while Teauama talked and did the cooking. Here is her story.

Mrs Teauama Bauro with her daughter in her house in Bikenibeu

Teauama gets up very early, before 6 am, and takes a bus to her employer's house. Her husband looks after the welfare of the children after she leaves home for the market. She has no time to have breakfast. It is important for all the vendors to arrive early to claim a share of fish to sell. There are 10 fishmongers working for the same owner but not every woman turns up every day; she may have other commitments. Selling takes place seven days a week, but for only a short period on Sunday because many go to church or take a rest on that day.

When there are plenty of fish, each woman takes as many as she thinks she can sell for the day. If fish are limited, they are evenly distributed among the 10 women, so that each can have some income for the day. If there is only one esky of fish to share, the employer will sometimes give the women $3 each or more plus an extra dollar for bus fares. Teauama thinks her employer feels sorry for them when there is not plenty of fish, or *e mate tari* (literally, the sea does not have life, which is to say the fish are not biting). There are also times when the employer gives the vendors extra cash, which he calls a bonus, for being in the market for too long.

The employer does not provide ice. If the fish goes bad, the blame is put on the vendors because they have been asked to take care of the fish by getting ice as needed. Because it is the responsibility of each seller to ensure that her fish stays in good condition, one woman may go by bus to buy ice cubes. This expense is recorded and repaid by the employer. Ice is not always used because the fish are cold, having been in a freezer the night before, and sometimes demand for fish is so high that the fish sell out in no time.

After collecting their fish at the owner's house, the fishmongers are transported on the company truck to the roadside at the central market in Bikenibeu. This is the only place where all of them sell their fish. In the past, the women used to spread out along the road but then, some years ago, the TUC (Tarawa Urban

Council) started to demand a fee of $5 per esky from each vendor. Because their employer advised his vendors not to pay, the TUC ordered them all to stop selling fish. When the women refused to stop, the TUC truck was sent out and the fishmongers were ordered to vacate their selling sites. This happened around lunch hour and the fishermen at the owner's house were surprised to see them returning early with their fish unsold. Following that incident, the employer rang up his lawyer and the lawyer discussed the matter with the TUC officials. Since then, a $5 charge per business has been agreed upon, on condition that all of the vendors for one company use only one location.

Each fishmonger earns 10% of her daily sales, which means that if Teauama sells $120 worth of fish in a day, she will receive $12. She uses this money for basic family needs. On occasions when *e maiu tari* (the sea is full of life, i.e. fish are biting) she can make sales of AU$100 to $120, and these may add up to more than $1,000 in one day.

At the market each woman calls to the customers, trying to sell her fish quickly so she can relax. On the day of our interview, Teauama had been the first to finish selling because, she explained, her esky was the first to be loaded in the morning and there were plenty of customers after a week-long shortage of fish. By around 2pm she had sold all her fish and then she waited for the others to finish.

When she wants to have lunch, Teauama leaves her fish on the roadside and runs home, as it is not far. The other fishmongers do not sell her fish for her because they all compete for customers.

The TUC market building which is near by does not have proper sanitary facilities and the women have a hard time when they want to relieve themselves. Most use the ocean beach as a toilet because the TUC toilet has been locked for a long time. Teauama feels sorry for the older ladies because they have to walk a long way.

Around 6pm, the truck comes to pick the vendors up and each of them counts her money with the owner's wife. As there are ten fishmongers, it takes quite a long time to do the counting. By the time everyone is finished and they get their pay, it may be 7:45pm or even later. Teauama gets home at about 8pm and starts cooking. On most days her husband helps because he is at home all day. Teauama is happy with what she is doing because, she says, she earns "enough for the day…We are lucky there are only five of us in the household so what I get is sufficient".

the reward of being able to support the daily needs of their family. Despite poor financial returns, the women said they keep working because they have no other choice and because at times the owner of the business gives them free fish.

To ply their trade the women require handcarts to move their ice boxes (eskies) and fish around. Some women own their own handcarts while others borrow one from their employer. The owners of better-developed businesses provide truck transport and ice cubes. The eskies belong to the business owners. Although ice is often used in the fishmongers' eskies, no ice is used on the fishing boats because the weight of the ice would lead to higher fuel consumption and thus higher costs (Ekueta, commercial fisherman).

WORKING CONDITIONS

Women's work in the fish trade is characterized by poor and unequal remuneration. Women in the informal fish-selling sector do not know the minimal pay that an employer is legally supposed to provide for their services. Some fish sellers have few or no options for earning income and therefore feel that they cannot make any demands on their employers. The women fear that, if they complain, the employer might select other women to do the job, as there are no signed contracts. There is even greater pressure on those women who are relatives of the business owner. In such a case, work conditions are never questioned and women choose to continue despite the small earnings they make for each long day of work.

Pay varies for the women and pay increases are rare. Even after one woman worked for the same employer for six years, she was not given any raise in pay. More than a third of the women interviewed were earning 10 cents per dollar of fish sold and two were given 20 cents per dollar of fish sold. In general, fishmongers operating in Bikenibeu and Bairiki areas were receiving 10 cents per dollar of sales while in the Betio area the rate had increased to 20 cents per dollar. Three of the women were earning a flat rate of between $10 and $20 a day. One woman did not receive her pay personally as it was included as part of her husband's income from fishing. In one other case, the employer habitually neglected to pay the woman at all, and in one case the pay scale varied (Graph 1).

Because daily earnings generally depend on the amount of fish sold, women are encouraged to maximize sales and to work long hours. On most occasions, the women report that they earn at least $10 per day and for this they work more than eight hours. The rate of remuneration is the same on weekends. While the hours are very long and the pay is poor, the women remain in the fish trade because they have few alternatives for making the money they need.

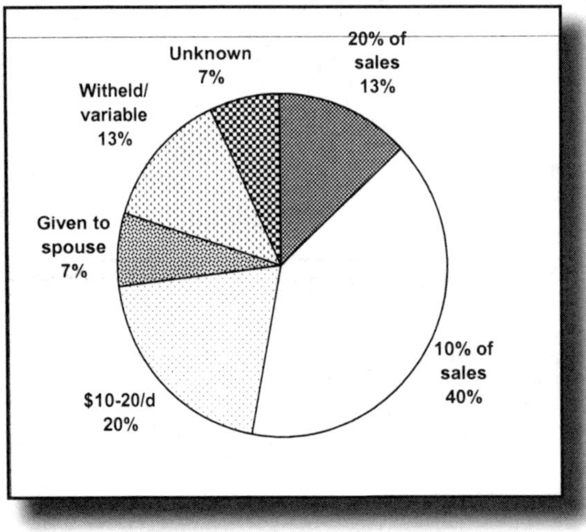

Graph 1. Arrangements for remuneration of female fishmongers on Tarawa.

Kiribati Provident Fund

The Kiribati Provident Fund (KPF) was set up in 1977 to cover wage employees. Both workers and employers contribute to the fund, which then provides certain long-term financial benefits to workers who experience unemployment, as well as some savings for their dependents should a worker meet an untimely death. Workers who are not automatically covered by the KPF may elect to voluntarily pay into the KPF scheme as a form of personal savings plan.

As explained by one small scale fish business owner, fisherfolk are not paid wages but instead are paid what is referred to as a bonus. Bonus payments, overtime and allowances are not regarded as wages, and therefore do not warrant payments to the Kiribati Provident Fund. However, according to the Acting General Manager of the KPF, it is the right of any employee to seek his / her worker's rights and if a fish worker laid a complaint against an employer, the employer could face legal charges. Some small scale fish businesses do put aside savings on behalf of the employees, which workers can collect at the time they decide to leave the employer. This is, however, not part of the KPF scheme. No security of work is guaranteed in the local fisheries trade and there are normally no signed contracts. At times when there is no fishing, fish sellers have no income.

Many people in the informal sector are not aware of their entitlements. In interviews with fifteen female fishmongers on South Tarawa, not one of them had knowledge of the Kiribati Provident Fund nor of the Voluntary KPF payment scheme. The respondents when interviewed were quite suspicious and not very willing to discuss issues concerning improvements in their welfare, especially in regard to the terms and conditions offered by employers. This reveals the women's insecurity concerning their livelihood. No one showed any interest in following up any problems with their employers. Most did, on the other hand, take more interest in the Voluntary Payment Scheme of the Kiribati Provident Fund (KPF), which does not involve the employer. According to KPF, these women need only to become registered as employees of any company before they can join the voluntary KPF scheme.

Regulations And Infrastructure Affecting The Fish Trade

The municipal councils demand fees from all fish traders but the fees are not uniform. In Bairiki and Bikenibeu, located within the Tarawa Urban Council (TUC), the fee that is charged for each business is $5 per day while in Betio the charge is $3 per day. While confused about these differential rates, the women interviewed have never asked why these charges are variable or even required, since their employer already pays an annual business license.

As we have seen, each of the three main markets have major flaws which limit their utility to women. Not one of the interviewed fish sellers is using the structures that are meant to be fish markets. The market facility provided at Bairiki is considered to be too distant from the public. In Betio, the fish sellers place themselves close to a bus stop where many people get off to attend bingo games. In Bikenibeu, the women line the roadsides where they have easy access to customers visiting the nearby bank

Mrs Bauro and a colleague sell candies and make handicrafts while waiting for an afternoon delivery of fish; the morning's lot of fish has sold out.

and supermarket. Good marketing skills and optimal siting are critical to success because earnings depend directly on the amount of sales the women make each day.

The fishmongers do not have basic facilities such as water, access to a toilet, ice storage, drainage and shelter. Some bring their own water but in general the lack of access to proper washing and drainage facilities is a major problem. In the late 1990s women fish traders were given several evacuation orders because they were blamed for draining the water they used onto the ground at the roadsides. In response, the fishmongers agreed to pour the water some distance away from where they sell. Given the general lack of facilities it is not known why payments to the council are required.

Education and Training Support

The women selling fish have had no access to training opportunities. They have little formal education and from observing their work practices it is apparent that they lack information on basic sanitation and proper handling of fresh fish. The fish sellers also need more information on post harvest fish processing so that they can preserve their fish when their employer distributes excess catch to his workers, as happens on some occasions. In all their years of working in the fish trade, only one woman interviewed had ever attended a fisheries training workshop. Some had learned some fish handling skills from visiting officers from the Ministry of Environment and Social Development. There have been several post harvest fish processing workshops and "start your business" workshops conducted in previous years by various governmental institutions, but except for one small workshop put on by SPC in 1997, these have never targeted the women in the informal fish trade.

WOMEN'S ROLE IN DECISION-MAKING

Women in Government Departments

In Kiribati, society is based on kinship and place and councils of older men have traditionally ruled. With independence, the parliamentary process has linked the traditional elites to the modern business elite in the capital (MacDonald 1996). Access for women to education came late in Kiribati, owing to the cultural belief that daughters need to stay in the home and be looked after. In recent years, as they gained greater access to education, some women have managed to secure high ranking positions in the public service. Yet, at decision-making levels, women's representation is still relatively minor. Education for women at the grassroots level and for those working in the informal sector is limited and many are not aware of their rights as individuals.

Rural woman salting fish in Aonopuaka village on Abaiang Atoll, north of Tarawa Atoll.

In the 1970s and early 1980s only males took up higher education in fisheries. However, since the late 1980s some women have entered fisheries studies and then become employees in the Ministry of Natural Resources and Development. Two female graduates with Masters degrees have worked in the Fisheries Department and one of these is currently undertaking postgraduate studies. In total there are seven males with Masters degrees employed by fisheries. In recent years, there has been a slight increase in the number of women gaining fisheries related degrees but two have not yet been posted to work. At the diploma level, there are one woman and four men employed. It can be anticipated that in the coming years there will be more women taking fisheries studies as more go into the science streams at school. For now, the percentage of women qualified and working in fisheries remains low.

The limited number of women in the Department of Fisheries is important because, owing to their experience, women may look at fisheries from a perspective that is different from men. Having women making decisions in the fisheries area could improve the chances of having issues related to women in the fisheries discussed. Also, women's presence in different work sectors enables research into areas that were ignored in the past. An assessment of needs of fish workers that is currently being done by the national, women's non-governmental organization, AMAK, is an example of a new area of work that may stimulate revision of some important council bylaws and other regulations affecting these female workers.

Another important issue where women's involvement in decision-making is important is the pollution of Tarawa Lagoon. Communities on South Tarawa still lack awareness of how to keep their inshore clean and women, if involved, could play a key role because this problem affects women directly and also, much of the pollution is the result of poor sanitary practices at the domestic level.

Representation in the Municipal Council

The Tarawa Urban Council (TUC), the decision making body controlling the fish trade, comprises seventeen men and only one woman. Thirteen of the council members represent the various villages and towns included in the Tarawa municipality. These seats are and always have been occupied by elected men. Three council members are the local members of parliament (all male) and one is a representative of elder men. Women elders have no representation. The one seat occupied by a woman is reserved for the representative of a women's non-governmental organization (NGO) that is affiliated with the council. It is not known exactly how much villagers use their councillors to voice their concerns. The same applies to the woman representative on the Tarawa Urban Council. There is a history of complaints and enquiries lodged at the national women's NGO office (AMAK), which suggests that women either are not aware that they have a representative on the Council that they should go to, or perhaps they are shy to approach her. The councillor has not been proactive in advertising her role as a women's representative.

Fishmongers at Bikenibeu waiting for customers.

In future it should be possible to improve communications between women and their NGO representative so as to have women's concerns heard. Also, AMAK has suggested that a businesswoman from the Chamber of Commerce would be an appropriate addition to the council. The Clerk of the TUC, Ms. Kirata, has recently stated "I see no reason why the idea of involving other relevant bodies may not be welcomed. The women are represented on the Council by a woman NGO member that is linked to the Tarawa Urban Council. It is through her that women can lodge their enquiries."

IMPROVING CONDITIONS FOR WOMEN IN THE FISH TRADE

This case study was undertaken as action research with a view to identifying the key issues affecting fish traders and taking action to improve their situation. The information compiled will be used to design further interventions by the national women's umbrella organization, AMAK.

Knowledge of women's roles in the fisheries needs to be extended, particularly to government bodies and NGOs that could help fish workers exercise their rights to food security and livelihood. In recognition of the importance of fisheries to women, fisheries issues have to become one of the priorities of the Kiribati National Council of Women. To continue to neglect these issues will further disadvantage women who are already marginalized.

Normally, Kiribati women think they have no legitimate say in decision making; men are still regarded as the natural decision-makers. But for women in the fish trade to improve their position they will need to be empowered to at least voice their concerns and seek assistance. Better representation of women on municipal councils and on other decision-making bodies is needed. This would help to bring neglected issues of fishmongers and other women working in the informal sector to the attention of those who have the power to assist.

One of the important findings of this study is that women often work for many years selling fish even though their hours are long and their remuneration is low. As a point of comparison, Kiribati office cleaners, who are mostly females with limited education, earned an annual salary of AU$3252 in 2003. This works out to $62.54 per five-day work week. In contrast to this, fishmongers work six or seven days a week. Work days are much longer: from early in the morning until after 7pm, and working conditions are both uncomfortable and unsanitary. On a very good day, fishmongers may earn $100-200 but most of the time their income is much less ($10-20) or even non-existent. The need to have cash on a daily basis and the dynamics of a small island society keep women silent. They are often working with extended family, relatives and neighbours, which makes it difficult for them to demand better wages and working conditions.

Regardless of what type of employment one has, a person should be provided with the regular entitlements of an employee. Women, who form a large labour force in informal economic sectors, need relevant information and training so that they can help themselves.

Women from all around the South Pacific Islands take advantage of training provided by the CETC in Fiji, which includes learning about marine resources.

Previous research conducted on Tarawa has shown that fish and food poisoning are relatively common on the atoll while standards of sanitation and hygiene and awareness of the dangers of poor handling of fish are particularly low on South Tarawa (Novaczek & Chamberlain 2000:9). Fishmongers are charged with the responsibility of maintaining the fish in good and healthful condition. This is difficult in Kiribati because not only is ice not readily available but also the women lack knowledge of safe post-harvest handling and sanitation. This is another training gap.

To better the conditions of female fish workers, the KNCW should advocate for training that will cover the gaps identified in this study. As a result of this action research, AMAK plans to locate funds to provide training for fishmongers. The intent is to improve market facilities and standards of hygiene. The first step will be the completion of a needs assessment funded by the Canada-South Pacific Ocean Development Program.

Others who could become involved in developing programs for fishmongers include the graduates from the Community Education Training Centre (CETC) based in Suva, Fiji. Women who have been through the CETC program are meant to share their new knowledge and skills when they return to their countries. Seafood handling and sanitation are among the subjects they learn, so they represent a human resource asset that could be tapped for program development in Kiribati.

Tony Chamberlain of the Marine Studies Program at the University of the South Pacific trains CETC students in how to handle fresh fish.

I have noted that, in addition to the lack of adequate representation, training and information, the facilities now offered to women for marketing are inadequate. Because women's work in handling and distributing fish is vitally important to local food security, they ought to be more effectively supported.

REFERENCES

Catala, R. 1957. Report on the Gilbert Islands: Some Aspects of Human Ecology. *Atoll Research Bulletin* 59:1-67

Handy, M. 1986. *Supporting Women's Enterprise Development.* Peace Corps Information Collection and Exchange.

Loumala 1984. Sharks and shark fishing in the culture of Gilberts, Micronesia. In B. Gunda (ed) *The Fishing Culture of the World, Studies in Ethnology, Cultural Ecology and Folklore* Vol.11:1203-1250

MacDonald, B. 1996. Governance and political process in Kiribati. *Economics Division Working Papers* 96/2, Research School of Pacific and Asian Studies. Canberra: Australian National University.

Ministry of Finance and Economic Planning 2000. *A Report on the 2000 Census.* Tarawa: Government of Kiribati.

Mitchell, M. J. 1994. *Subsistence, Surplus and Seaweed: Women's Fishing and Marine Work in Kiribati.* Unpublished report, 174p.

Mulik, J. & Werner, L. J. 2002. *Time Use Survey for Gilbert Island Group*. Ministry of Environment and Social Development, Tarawa.

Novaczek, I. & A. Chamberlain 2000. Needs assessment and strategic approach for the post-harvest fisheries development project. *Marine Studies Technical Report* 00/1, University of the South Pacific. 36p.

PACFAW 2002. *Kiribati Status of Women Report*. Suva, Fiji.

Tekinatiti, T. & V. Wichman 1995. Materials from focus group meeting, Fisheries Division, representatives of fisherwomen, and Atoll Research Programmes/ USP research staff, Sept 1995. *SPC Report.*

UNESCO 2001. *Assessment of Best Practices, Resources and Gaps in Gender, Science and Technology*. Samoa.

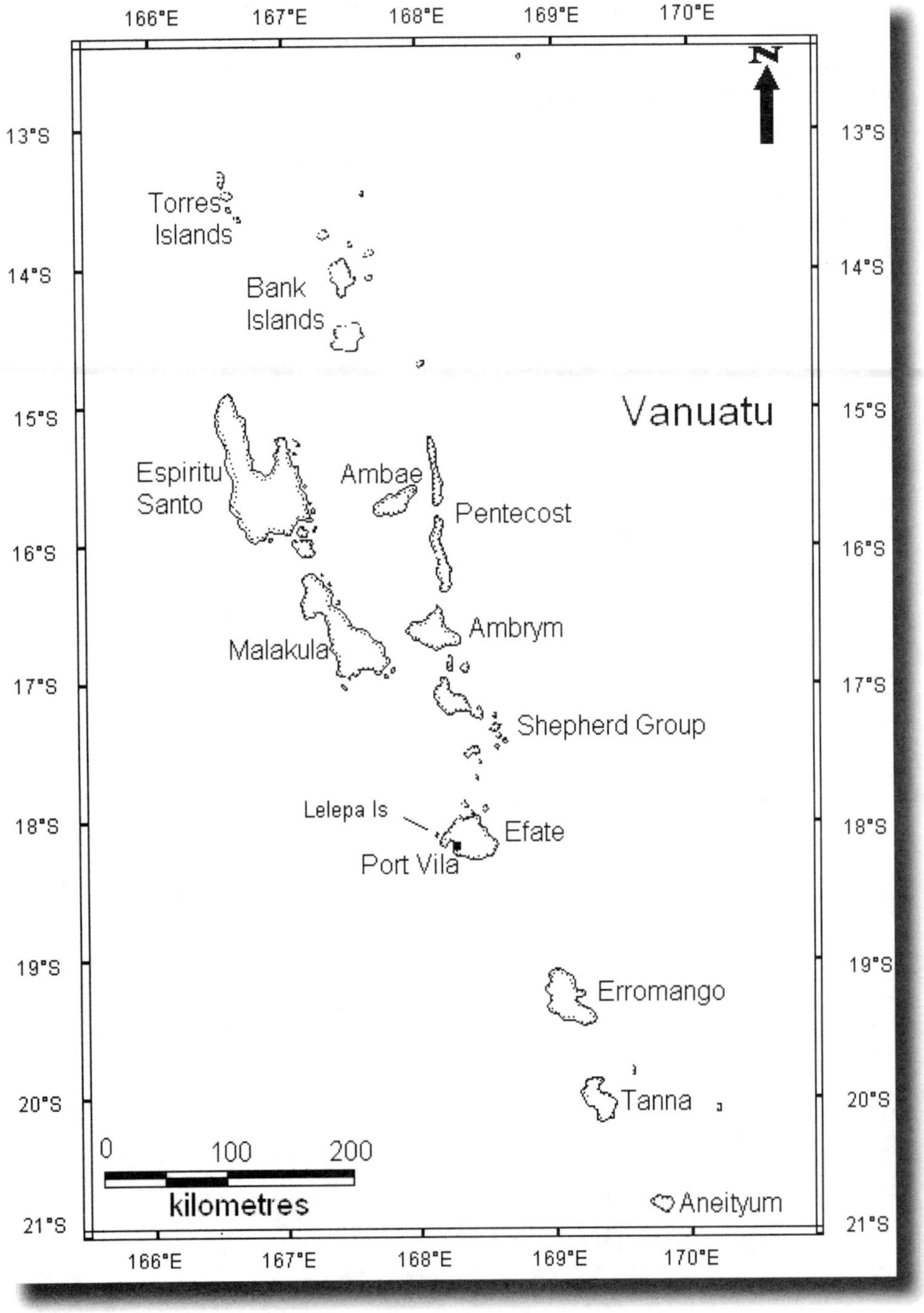

Map 1. Vanuatu, showing Lelepa Island lying off of northwestern Efate.

13 Gender, Generational Perceptions And Community Fisheries Management In Lelepa, North Efate, Vanuatu

Jean Tarisesei and Irené Novaczek

"When I was a boy, people fished only for food, not for vatu. You could spend just a small time on the stone wall on the beach and see plenty fish, including big ones.... In those days we fished only with a spear."

— *Kaltoa Monalapa, age 80*

"We need fish for the future of our children ... Their fathers took all the big ones as well as small ones. Then the chief made a ban. He said, 'Stop. One day the children will grow up and ask what is fish? What is crab and shell? Because their fathers will have killed every one.' So the chief made a ban on the island and the fish are now becoming more abundant."

— *Jaklin Ropi, age 35*

THE CHANGING FISHERIES OF VANUATU

In recent decades and especially since national independence in 1980, there has been significant pressure on rural Vanuatu to develop and join the cash economy. As a consequence, people have come to think more of harvesting marine resources for commercial purposes. Those having access to urban centres now transport their catch to sell in public markets. The emphasis has shifted from communal sharing of resources to individual efforts in harvesting, processing and selling of fish. Many communities also allow outside interests such as the aquarium trade to harvest corals, clams and small fish from their reefs in return for access fees.

This case study focuses on the subsistence and small commercial fishing activities on Lelepa Island, north Efate, Vanuatu. On this island, fisheries have gone from abundance to decline. The need for resource management has been recognized and acted upon. Yet, finding the balance among subsistence needs, economic development and resource conservation is difficult and contentious. In this case study we will document the history of change in the island's fisheries and explore how cultural practices have influenced marine resource use, management approaches and compliance or resistance to fisheries rules on Lelepa. In particular we will look at culturally defined gender roles and traditional power relationships between genders and generations.

CASE STUDY METHODS

Vanuatu Cultural Centre (VCC) researchers collected some data on traditional harvesting and management practices in 1999-2001. Additional research in November 2003 used village observation and semi-structured interviews with key informants, including males and females ranging in age from 12 to 80 years. All information concerning the history and conditions in Lelepa that is reported in this case study was derived from interviews with people of Lelepa, and where possible, verified through cross-checking facts from three or more informants. Information and opinions particular to one informant have been indicated by reference to the informant's name (in brackets) and, if relevant, their age and gender.

Woman returning from her forest garden on Lelepa Island, with a load of firewood

LELEPA VILLAGE

Lelepa is a tiny island on the northwestern side of the island of Efate in central Vanuatu. From Lelepa a short ride to Efate by canoe (the most common local transport) or motorboat (locally owned public transportation) followed by a half-hour drive in the back of a truck will bring you to the capital of the country, Port Vila.

Lelepa has seven small villages strung together on a narrow strip of flat land facing Efate. Behind them is a steep bluff covered in scrub and small trees, which leads to flatter high land suitable for gardens. In all the villages together there are perhaps 80 households, and 350 people, 41% of whom are under the age of 15. Only 8% are elders over age 60 (National Statistics Office 2000). Despite being so close to Efate, Lelepa people speak their own distinct dialect.

View of Lelepa Island from the sea.

It is said that two or three new houses are added to the villages every year. The population, however, does not grow as rapidly as it could because most young women marry into communities on other islands.

The village has its own school, several shops and a small medical clinic. There is a community hall and a women's centre where church groups carry out sewing, weaving and other income generating activities. Most people share a common religion, being Presbyterians. Because men over age 18 and virtually every girl and woman on the island belong to church groups, the church is central to village life.

Lelepa people originally came from the mainland of Efate during a period of warfare among villages. Because of these historical origins, Lelepa owns extensive lands on the Efate coast (Map 2). After the church was established on Lelepa in 1898, people from many islands and villages came to live beside it, giving rise to the seven villages here today.

Local Governance

Each clan that came to settle on Lelepa had a chief and the descendents of those chiefs are recognized today as small chiefs. The position of chief is inherited from father to eldest son or, if he is deemed unsuitable, the title may pass to a younger son, daughter or nephew. There has been only one female chief on Lelepa, and she acted as a caretaker of the position while her baby son grew old enough to take over (Kaltoa Monalapa, age 80). Currently there are 12 small chiefs on the island, and they recognize the authority of one paramount chief whose position is inherited.

Paramount Chief of Lelepa, Natamatsaru Taua

Small chiefs on Lelepa do not act in isolation to govern their own clan. They meet together with the paramount chief to make decisions affecting the entire island. This is called the Council of Chiefs. The paramount chief has an assistant chief to help him with administration. He also nominates certain men and women to act as customary police. They look after violations in local law such as theft and violent behaviour.

Although the paramount chief may make certain decisions on his own, he is expected to consult the Council on important matters. Council meetings may be called at any time, and are led by a chairman who is not a chief and may be male or female. Meetings and other council activities are funded by various resource access fees and by the fines levied on people who break customary laws. This money may also be used to address community problems. At Christmas the paramount chief holds a small party for the council and the police, where gifts such as woven mats and small amounts of money are handed out. The chief also provides a written account of how the village money has been spent (Kaltoa Monalapa).

VILLAGE LIVELIHOODS

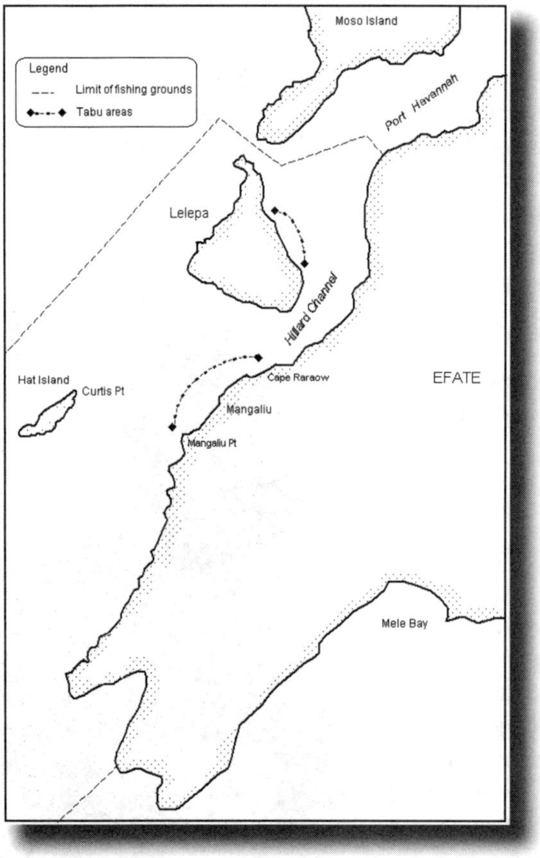

Map 2. Lelepa land and marine territory, showing the locations of protected areas of Lelepa and Mangaliu.

The land base of Lelepa is owned by individual families, who may sell it if they wish. Most families have a small garden on Lelepa as well as larger garden areas on Efate. As there is no stream or spring on the island, people depend on rain for their drinking water. Women carry clothes to the mainland for washing and when rainwater tanks run low, people must also carry drinking water back from Efate.

On Lelepa people subsist on their garden crops, shellfish, crabs and fish. Although there are some chickens and pigs raised for food, these are meat for very special occasions. Marine products are the primary source of daily protein. In addition to local food, people require cash for goods such as sugar, tea, kerosene, soap, tools and clothes. Cash is also needed for school fees and church contributions. In the past, cash was earned by processing copra but since this industry collapsed, cash income is derived from sales

A WOMAN'S LIFE ON LELEPA

Ann Marie Ken, a young woman of Lelepa, typically tends her garden every day. Some days she takes a canoe to the mainland to wash clothes in the river. Collecting firewood in the hills behind the village takes half a day. Ann Marie weaves mats and baskets in her weekly women's group meeting and also at home. To go fishing she canoes around the island to hand line for reef fish at night when the moon is shining. Sometimes Ann Marie goes fishing alone; other times with her husband or a girlfriend. If she does not own her own canoe, a woman must borrow one from a father or husband and so can fish only when the men are not using it.

Ann Marie Ken displays a catch of reef fishes

Ann Marie would like to learn how to use a spear gun but the men in her family refuse to let her practise; they say that a woman would blunt the tip of the spear on a stone, or would be too weak to load the spear.

Often, Ann Marie dives for giant clams, snails, chitons and octopus on the back side of Lelepa, on Hat Island or on the mainland. Some are for family food; some are sold in the Port Vila market. Young women such as Ann Marie often do not sell their shells themselves, but send them to market with an older woman.

of coconut crabs, land crabs, garden vegetables and marine products. Some women also sell handcrafted mats, baskets, sewn garments and shell necklaces. A few people, mostly men, hold office jobs in Port Vila or do construction work.

Tourism is an emerging economic driver in the area. Two families own boats that take tourists fishing, snorkeling, diving and whale or dolphin watching. This employs a number of young men who act as guides and boat crew. Outside tourism operators wanting to bring people into Lelepa waters are asked to pay an access fee to the paramount chief.

Above: Forest garden on Lelepa at yam planting time.

Right: Yams drying out to get them ready for planting.

Below: Tobit Jeffry and Lili Toni take goods to the city market early on Wednesday morning and will return the following afternoon.

THE HISTORY OF THE LELEPA FISHERY

Early 20th Century

Kaltoa Monalapa, an elder fisherman of Lelepa.

Elders on Lelepa remember when there were few people and many, many fish. The fish were large and tame. You could see them while standing on the beach and men caught them easily using hand held spears. A man bringing fish home in those days was expected to share his catch. No fish was ever sold; it was only given. "When I was a spear fisherman mostly I would go out fishing on Saturday to catch fish for Sunday. When I came to shore, mothers would be sitting there so I would give them some fish. Then I would … share a plate to each neighbouring house" (Kaltoa Monalapa, age 80).

Changing Gender Roles and Technologies

Women did not use a spear; they fished with bare hands or sometimes used plant poisons. Poisonous leaves could be thrust into holes in the reef by a woman diver, or placed in tidal pools during the big full moon tides of May through July. This would stun or kill any fish, making them easy to collect. Lelepa elders reported that gender roles in the fishery were firm, and only women collected *Trochus* and other shells for food.

As *Trochus* became commercially valuable, men started to collect them. At first the fishing pressure was seasonal. "In December they would go out and collect a lot and bury them. Then at Christmas the shells would be cleaned up and the traders would come and they would be sold. After Christmas they would forget about the *Trochus* and not collect them for a while. There were not so many people and *Trochus* collectors worked a small area; the rest was untouched so the shells had time to grow" (Kaltoa

HOW TO KEEP FISH TAME

You have to kind of look away…when the big crowd of fish come ashore you must not catch them because when you spear them, every one will learn what the spear is. When all of them are going back again, one of them will stay behind – you catch that one. When you do it like this the fish are tame. It's the custom way."

— Kaltang Zakaria, age 55.

Kaltang Zakaria, small chief and former fish trader on Lelepa.

TRADITIONAL OCTOPUS FISHING, AS TOLD BY THE ELDERS OF LELEPA

When a woman went to the reef she would carry a branch of a particular tree called *natongtong*. Holding the branch in her hand she would sing the traditional octopus calling song while walking over the reef. Any octopus within range would emerge from its hole in the reef to be picked up by the woman, swiftly killed and placed in her basket.

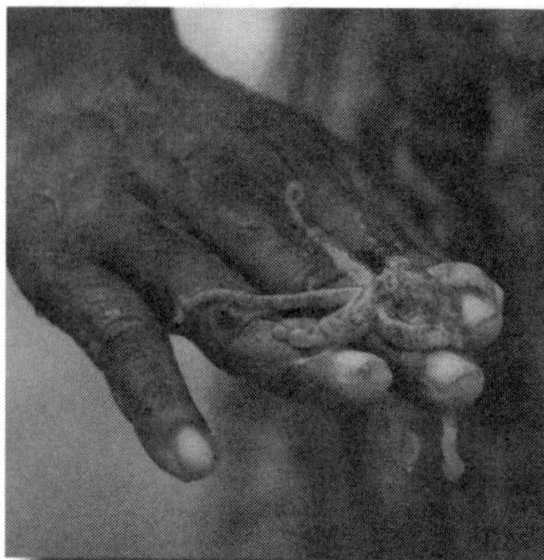

Baby octopus.

Monalapa, elder). For many years these valuable shells were taken to Port Vila on a ship owned by a Lelepa man who also traded in copra and brought back dry goods to sell in the villages.

Over time, the harvesting intensity increased and spread over wider areas. "If today the chief opened the season for *Trochus* the sea would be white with people collecting – every man, woman, girl and boy. They would go to every place – there is no space left for the *Trochus* to hide" (Kaltoa Monalapa, elder). Village women who are now in their 50s and 60s say that the entry of men into the *Trochus* shell fishery and the increase in men's harvesting of shellfish for food have contributed to a decline in abundance of shells. Men collect a variety of shells to eat while on extended fishing trips.

As time went by, technologies and gender roles continued to change. The idea of throwing spears from a canoe was introduced by a schoolteacher from another island (Kaltoa Monalapa). Early versions of spear guns were made, consisting of two sticks joined by a rubber band, with a point of wire on the end of one stick. By the 1940s fishing lines and hooks had been introduced and this new technology was adopted by women as well as men. Newer, more effective spear guns also began to be used and traditions that kept fish tame were abandoned. "Instead of waiting for the fish to come to them the men chased the fish into deep waters. The use of spear guns like this made fish wild" (Kaltoa Monalapa, elder). The next innovation, to use underwater torches and spear fish as they slept on the reef at night, is also regretted by elder fishermen.

Nets were introduced in the 1970s and these, like spear guns, remained exclusively male technologies. At first the fishermen allowed the nets to hang passively but then they changed their technique and began to drive fish into the nets. People believe that net fishing caused some

The beach in front of the villages on Lelepa is piled with coral rubble.

fish such as *naewo* (*Scarus* sp.) to become scarce. "*Naewo* could be attracted into shallow water by dangling hermit crab meat on a weighted string. These big fish swim in schools and used to be plentiful. They are beautiful and shiny and taste good. In the old days people speared them; now they use nets and this has fished them out" (Kaltoa Monalapa, elder).

Over time fish habitats were degraded through both natural disasters (earthquakes and cyclones) and by human activities. The reef in front of the villages on Lelepa is heavily damaged and the beach is heaped with coral rubble. This area, where canoes are landed, has been trampled and exhausted by decades of intensive fishing and gathering.

Commercialization and Further Resource Decline

A reel called a byskel is locally made and used for deep water fishing from a canoe.

In the 1970s Lelepa people began to sell fish in Port Vila. Women sold cooked fish, octopus and giant clams on *laplap* (traditional puddings made from coconut mixed with cassava and other root crops). Men sold fresh fish. The elders say that because of the increased pressure to catch fish for sale, by the end of the 1970s Lelepa's marine resources were already in decline.

The 1980s brought national independence and a new emphasis on development of export fisheries. Motor boats with reels for fishing red snapper at 100 – 600 m depth came into use. A home made reel called a *byskel* was developed for deep water fishing from a canoe. Once again, it was men who acquired the new technology.

The local fish trader who bought reef and deep-water fish in Lelepa from 1986 to 2002 saw evidence of further decline through the

MAKING TRADITIONAL LAPLAP

Right: Elder Leisny Ruben (foreground) grates taro while Karolin Fred prepares coconut cream to mix with it.

Below: Karolin places the taro and coconut mixture, wrapped in leaves, to cook on hot stones in a ground oven. The result is called laplap.

Above: Laplap may be sold with fish, meat or vegetables on top. It is one of the most popular traditional fast foods sold in the Port Vila market.

Left: Leiatu Billy from Mangaliu village sells laplap in the city market.

1990s. Whereas he could buy up to 120 kg of fish a day from Lelepa fishermen in the late 1980s, ten years later he would be lucky to go to market with 30 or 40 kg, even though the number of fishermen was the same (Kaltang Zakaria, former fish trader).

Current Fishing and Fish Marketing Activities

Women continue to provide family food, primarily by diving for shellfish and octopus. Although some shellfish are exclusively food items, others are sold. Some shells, such as cowries, are left to rot so that the intact shells can be sold to tourists. The most valuable shells, green snail and *Trochus,* have been overfished. Green snails are so rare they may be locally extinct, and *Trochus* are uncommon. The next most valuable shellfish is giant clam. The name seems ironic; clams are rarely allowed to grow to their normal adult size before being harvested. Women report that some of them have given up selling shells in the market because they now can find only enough to cover food needs.

Shells for sale in Port Vila

Both men and women target reef fish during the day but the preferred time for line fishing on the reef is at night when the moon is shining. Men spear fish at night using torches. They scour reefs around Lelepa and Hat Islands as well as along the Efate shore, excluding only two small protected areas in front of the villages of Lelepa and Mangaliu (Map 2). The smallest fish end up as family food. Fishers report that many of the big fish, especially wrasse and grouper, are now rare.

Douglas Kalotiti, field worker for Vanuatu Cultural Centre, displays crabs caught by women and heading for the city market.

Only men fish in deep water with a reel, but on occasion women help to paddle the canoe out to the fishing ground. Red snapper fetch a good price in the commercial market and are generally sold, not eaten at home. Fishers report that they can no longer catch large red snapper close to shore but must fish in ever-deeper water.

The current fish buyer in Lelepa sells fresh fish to a Port Vila supermarket and this market takes whatever amount of fish he can obtain. The supermarket provides free ice and bait (imported sardines). In addition to a guaranteed market and immediate pay, Lelepa fishermen will be sought by the buyer's motor boat if their canoe fails to return to the village at the expected time, and they can hitch their canoe to the motor boat if wind and waves make the return difficult. The buyer also collects fish from canoes during the day to put them on ice and keep them in top condition. Thus, Lelepa fishermen are encouraged to maximize their fish catch, and the bulk of fish caught goes to the market rather than to tables in local homes.

The Aquarium Trade

In recent years the aquarium trade has entered the Lelepa fishery. Several foreign owned companies are active in Vanuatu, where they collect small fish, live corals and juvenile giant clams for the North American and European markets. They may employ local divers or bring in workers from the Philippines. In the past, the companies paid one fee of 10,000 vatu (US $90.70[1]) per month to the paramount chief to gain access to the entire Lelepa fishing ground. The small chiefs objected to this arrangement. Now, companies must pay each small chief when they want to fish in front of his particular shore. Once an access agreement is signed, no one monitors what or how much the divers take or what technologies are used. "If they sign a contract then the land owner leaves them free to take what they like" (Meto Kalotiti, chief of Hat Island). The trouble with this became obvious to the chief of Hat Island when tourists who had paid to dive on the reefs complained about the disappearance of brightly coloured small fishes. In this case, the chief must choose between the aquarium trade and ecotourism; they cannot coexist.

Definitive scientific information on the impact of aquarium fisheries is lacking (Hughes & Connell 1999) and the possibility that these activities will have long term, negative impacts on reefs (Pyle 1993, Sauni et al, this volume) cause concern. Several elders in Lelepa said that they would prefer to ban the aquarium trade from the fishing grounds. The paramount chief claims that the Council of Chiefs has been looking for some strategy to control the activity more effectively but they lack information and expertise and "Department of Fisheries don't help much". It is generally perceived among Lelepa's leaders that the government is most interested in promoting the aquarium trade to generate foreign exchange.

Bêche de Mer Harvesting

Bêche de mer (sea cucumbers) are delicacies in the Asian market but in Lelepa they were not recognized as edible until the late 1970s. At that time a harvest was arranged to raise money to renovate the church. "Someone came here to show us how to process them. We cooked them and dried them on the beach on the other side of the island, and put them in sacks... Everyone went out collecting and sold them to one buyer" (Kaltoa Monalapa, elder). After the sea cucumbers had been fished out, the buyer did not come any more. In the late 1990s a second harvest was attempted but village people said that taking out the sea cucumbers affected the quality of the water. They feared that the frequency of fish poisoning would increase. A British biologist working with Department of Fisheries, backed up by a field worker from Vanuatu Cultural Centre, also advised the paramount chief to limit the harvest and so it was called off after about three months.

THE EVOLUTION OF LOCAL MANAGEMENT PRACTICES

Traditional Tenure and Limits to Fishing

Because of their extensive land holdings, Lelepa people control significant marine resources (Map 2) around their island and in front of their land on Efate. Together with the neighbouring village of Mangaliu, they share the right to fish around Hat Island. Lelepa fishermen claim exclusive rights to these fishing grounds and will chase away outsiders. Alternatively, chiefs will demand access fees from commercial enterprises targeting their marine resources. Thus, traditional tenure restricts access and controls fishing effort to some degree.

As noted above, gender roles in the fishery also limit fishing pressure. Cultural expectations and the weight of tradition limit the access of women to the more powerful fishing technologies. Although women have entered some fin fisheries they do not engage in the deep-sea fishery, nor are they encouraged to use spear guns or nets. These are generally perceived as more dangerous and masculine fishing activities. Women are expected to prepare food and care for children and elders. Women take pride in these responsibilities, which consume much time and require them to stay close to the home. In this context it makes sense for fishing tasks and technologies to be divided along gender lines. Motor boats, nets, reels and speargunѕ are also too expensive for many rural women to buy. As a result, although many women fish, their fishing power is limited. "Fishers who sell to me are both men and women… Many women fish but they do not catch so much" (Douglas Kalotiti, Lelepa fish trader). Although these gender roles may make it difficult for women to advance economically, there is also a benefit. The limitation on female fishing power may well have saved some fisheries from even more dramatic decline.

Lelepa women Hannah Jimmy, Salome Max and Henia Billy sit in the shade with their children and work on handicrafts.

As noted above, in spite of the limits to fishing that exist, resources have been declining noticeably since at least the 1970s. By the mid 1990s, Lelepa village faced a crisis. Demand for fish and shellfish had outstripped supply and the excess available for sale had shrunk dramatically.

Historical Precedent: The *Trochus* Harvest Bans

In the distant past, according to the elders, there were no controls placed on fishing because people were few and fish was only for food, not for sale (Kaltoa Monalapa, age 80; Toumel Leitapang, age 74). From about the 1960s, the paramount chief controlled the commercial *Trochus* harvest by placing temporary harvest bans on the reefs (Natamatsaru Taua, Paramount Chief). When the ban was lifted, generally near Christmas time, divers were allowed to take only the larger shells (Steven Marova, elder).

In the early 1990s the paramount chief, inspired by these historical bans, decided that *Trochus* and green snail stocks were in such perilous condition that all collecting should be banned throughout the Lelepa fishing grounds. Today, anyone taking *Trochus* should be fined but poachers in motor boats come from other communities and Lelepa men and women also freely admit to violating the ban. Some claim to not know that a ban exists. Although some villagers report violations to the chief, enforcement is weak. "Sometimes he will do something about it but other times he does not" (Lelepa woman).

The Lelepa Reef Protected Area

After setting in place the ban on green snail and *Trochus* the paramount chief decided that a marine protected area (locally called a banning area) was needed to conserve fish. He credits the previous chief for the idea. "The previous chief didn't think of a permanent ban but he did think we needed

Traditional fishing canoe on Lelepa beach.

to preserve the marine resources and not let them die out. So when I became chief we made a village law banning the area. The leaders of the community helped make the decision – the small chiefs and church leaders " (Natamatsaru Taua, Paramount Chief).

Elder Stephen Marova of Lelepa remembers when the fish were plentiful and tame.

In fact, a critical mass of opinion had been building for some time. In the 1980s Kaltang Zakaria, who later became chairman of the Lelepa Council of Chiefs, attended an environmental conference in Fiji. He brought back information about community based resource management and began to urge Lelepa chiefs to take action. Government departments of fisheries and environment reinforced the message through radio programs that urged Vanuatu people to be environmentally conscious. A third influence came from the Vanuatu Cultural Centre, whose field workers have a mandate to promote environmental responsibility. When a fisheries biologist came to the island to do research, that seems to have been a tipping point. "The chief got the idea from an English biologist… he said that in the breeding time the fish lay eggs and when they hatch they come to Lelepa and shelter here because the water is alive. He said … the paramount chief should have a ban on some area so fish could be safe... First he told the paramount chief, then all the chiefs met to talk about it. When the chiefs had made their decision they called all the people together to tell them why they would put the ban in place" (Truti Kalotiti, age 30).

The elders of Lelepa also exerted considerable influence on the debate. "It was the old people who talked about the change in the reef and started the move to make a banning area" (Kaltoa Monalapa, elder). "The elders were saying: Look at the change in our reefs. Before you could easily spear fish – now no more. The elders talked to people and then asked the chief to act. The chief asked fisheries to come in and talk about conservation… and then they decided to go for a banning area" (Annie Kalsong, elder).

The chiefs decided that the best location for a protected area would be the degraded reef directly in front of the villages, because everyone could help monitor it. They prohibited line fishing from shore, gill netting, and spear fishing and diving to collect shells along 2 km of shore and extending from the beach to the reef crest. After some heated debate it was sorted out that line fishing from a canoe just outside of the reef crest would be permitted, even though this violates the spirit of the ban and removes some fish that are supposed to be protected. This weakness is of concern to some people but is welcomed especially by women with small children who find it convenient to fish close to home.

Now that the ban has been in place for more than 6 years, people report that fish and shellfish, including *Trochus* and giant clams, are increasing in the protected area. According to the elders, the resources are still much less than they were long ago, but there has been an improvement in fin fishing all around the edges of the protected area. So far there is no sign of any spillover effect on shellfish resources.

Dissenting Voices

Many fishermen objected to the idea of a ban. "Fishermen attacked the chief... They said 'You try to stop us fishing but we help you to develop. Do you feed the fish? Did you create them?'" (Kaltoa Monalapa, elder). According to women, the men were upset because for them, the fishing area in front of the village was the easiest place to fish to make money to buy kava and cigarettes.

Young men also spoke up at the community meeting but their concerns were dismissed. "We asked the chief not to ban it but he did anyway. The chief said that when we grow up and marry, our children will need to have a banning area. That did not make sense to me" (Young man age 20, Lelepa).

For women the problem had other dimensions. Once the ban was in place, women in the centre of the villages had to travel up to 1 km before they could gather shells for food and small income. This meant that for many elderly women, their fishing days were over and they had to start depending more on their children and grandchildren. Elderly women also shouldered added childcare responsibilities when young mothers went fishing. Although some women at first resisted the idea of a ban, "There were other women who tried to explain why it was important. At first some women found it hard to understand. But once they had it explained so that they could understand, then they agreed" (Truti Kalotiti, age 30). In general, women found the chief's argument compelling and were concerned for the future of their children. The sacrifices they have made are an inconvenience but there are still many fishing areas open to them. Most women have the use of a canoe and so can paddle to the more distant grounds. Discontent may still flare when there is a special village occasion and the chief tells village men to gather meat for a feast. Some women whose husbands are not fishermen complain that they have no place to gather meat for visitors.

When interviewed, village men claimed that the protected area was their idea. They noted that women had neither expressed concern about resources nor openly favored the ban when it was proposed. The women agreed that this was so because until the elders started to talk about it, they did not realise how badly the marine resources had been degraded. As for the village meeting, women said nothing because "in this village they follow custom. They are afraid to talk in front of men... if there is a man around they will not speak out, they just think" (Truti Kalotiti, age 30). Some women also felt that, if they had spoken up in favour of the ban, that might have made the men reject the idea.

The majority of men interviewed voiced the opinion that women did not care about the future of the fisheries. The entry of women into commercial fisheries was characterized as greedy and not commended as an effort to support families. Women were blamed for breaking the *Trochus* ban and for collecting too many shells, even though men are also active shell harvesters. One man went so far as to describe women as "mad animals". "They don't respect the chief. They want to go by their own and take whatever they want". This gender stereotype may reflect the fact that women's fishing power and their economic options are limited, yet they are expected to provide food for their families. This leads women to try to maximize fishing effort when they have any opportunity. Edible shells and fish are collected regardless of size. But when asked why they need money, women always say: for family food, school fees and church contributions. They are not rampant consumers, nor do they lack understanding of the need for conservation.

These days, women play an important and unacknowledged role in educating their young children about the protected area. "The children know they must not catch fish in the banning area. Their mothers teach them and they understand. When they go to the sea they look at the shells but they do not pick them" (Jaklin Ropi, age 35). "The small children go to the reef and if they bring some shells up to the house the women make them take them back" (Leisny Ruben, age 73). A boy of twelve, who admitted

Lelepa men work to clean and chop kava roots that will be squeezed to prepare a potent drink sold in the evening at the local kava shop, or nakamal.

to sometimes fishing in the protected area, was asked what he would do to conserve resources if he were chief. His immediate response was that he would "keep the banning area so that there will still be fish in the future".

Opening the Protected Area

Although in theory there is a permanent ban, there are occasions when the paramount chief opens the protected area. If the ban is lifted only long enough to collect food for a feast, village people do not object. Occasionally, near Christmas time, if *Trochus* in the protected area are deemed to be large and abundant, the chief lifts the ban on *Trochus* only and allows each diver to take one bag full of shells.

Much more contentious is when the chief allows outsiders access in return for a fee. In the late 1990s, in exchange for money, he allowed the aquarium trade divers to harvest small blue fish from the protected area. "The community met to talk about it and they asked the chief to stop it, so he did" (Annie Kalsong, elder). Several years later the small chiefs sent divers into the protected area to take clams to sell to the aquarium trade. "The people felt sorry when they saw that the baby clams had been chiseled out of the reef" (Annie Kalsong) and indignation grew after people learned that most of the clams died from rough handling. "The people attacked the chief for that. It was the members of the Council of Chiefs who were taking the baby clams. They did not consult with the village people first; they just took the clams and sold them. The people said: 'You put the ban on us and then you broke the ban yourself!' The people were cross… They also said that the paramount chief knew what the small chiefs were doing and he should punish them" (woman, age 30). For many young men this was the point at which the Council of Chiefs lost their legitimacy. The perception that the chiefs are willing to break their own rules for personal financial gain has led to problems with compliance in the protected area. Also, young people suffer from a lack of access to information. Whereas elders consider the chief to be honest and legitimate, and can see from his financial statements that he is using village money for good purposes, younger people do not have access to such reports and this leads to suspicion and distrust.

Compliance and Enforcement

Every year some people are brought in front of the Council of Chiefs to be punished for fishing in the protected area. Mostly they encroach around the edges where there are fewer eyes watching. Non-compliance was worst in the early years of the ban but declined after the police explained that under the constitution, custom laws are accepted as state laws and can be enforced by government authorities (Douglas Kalotiti, VCC field worker[2]).

Occasionally the poachers are women who, the chief explained, do not re-offend after being fined. Some men, particularly the younger ones, do re-offend after being fined. For them, the protected area is a symbol of chiefly authority and breaking the ban is their way of voicing objections to the paramount chief's decisions. It is easy for the young men to fish in other places where the fishing is better but still they enter the protected area, especially when the paramount chief is off the island. They do not fear the small chiefs because they know that they also have stolen *Trochus* and clams from the protected area. It seems that in general, other villagers, including the custom police, have given up trying to enforce the ban on certain strong willed young men. "The police gave up trying to stop me.... My mother stays quiet; she can't do anything.... The elders can't say anything either" (young man, age 20).

A church committee takes a rest from their community activities.

FUTURE MANAGEMENT OPTIONS

The shells and fish in the protected area show signs of recovery but the remainder of the Lelepa fishing ground continues to decline. When asked what options they have for management of the fisheries, most people could only suggest extending the ban on fishing to other areas. But, as one woman noted, "There could be a ban on fish or on shell, but not both. People here get most of their meat from the sea. There can't be bans everywhere or people will suffer. They will have to go and steal from other people's fishing areas. You have to just ban some, and leave some open" (Truti Kalotiti, age 30).

People depend so heavily on marine resources, the extension of protected areas could cause real hardship, and enforcement would be very difficult especially against fishermen from other communities. Various people suggested that the chief could put notices on the radio to warn outsiders to stay out of Lelepa's waters. This was tried once before for Hat Island, but it was ineffective. For his part, the paramount chief would like to see more active government support for the enforcement of traditional bans.

A young man suggested reducing access to outsiders. Currently any person related by marriage to a Lelepa family has the right to fish on Lelepa grounds while visiting that family. The suggestion was to limit this particular right so that only current residents have access.

Women suggested that a ban could be put on shells because women would be able to focus on gardening or handicrafts to raise money. Such a ban would have to come from the chief. Women could not imagine developing a management strategy of their own and doing it voluntarily. They also feared that the men would never accept any such ban. A variation on this was the suggestion that seasonal bans be employed so that people would be encouraged to focus on their gardens during the months when that is most productive, and to go shell fishing in months that had the largest full moon tides (April – July).

Motorboat owned by Douglas Kalotiti serves as public transport as well as for commercial fishing.

With the ongoing increase in population and development of the cash economy in rural areas, the number of village-based marine resource management measures implemented by community leaders in Vanuatu more than doubled between 1993 and 2001 (Hickey & Johannes 2002). In some parts of Vanuatu, chiefs have banned both night fishing with spear guns and the use of gill nets (Johannes 1998). In Lelepa the chiefs also considered this approach, but were met with resistance from fishers who had invested in these relatively costly technologies. "They made the suggestion to ban the gill net and spear guns; but ... most of us got spear guns and for me, I've got 10 spear guns on my boat. And the last time I said if you want to ban me, please could you refund me back all my money? Then they didn't go to that solution" (Douglas Kalotiti, fisherman).

Elders bemoan the change in cultural values over time and suggest that resources could be conserved if people shared more and reduced their need for cash. A return to traditional ways does not, however, seem likely in light of the desire of people to enter the cash economy. The paramount chief suggests that young people reduce their need for cash by building traditional style houses. Yet, all the new houses going up on Lelepa are made of concrete.

Whatever the Lelepa people decide to do to secure their future, legitimate leadership, the equitable sharing of benefits and sacrifices, and public awareness are keys to success. Women represent a largely untapped power that could become a force for conservation and management. The potential role of youth is also neglected and young men appear powerless to do anything except engage in destructive activities as a form of political protest. Because they lack access to information, youth also may develop uninformed opinions about the use of village funds. They are particularly sensitive to perceived inequities in the distribution of benefits from the fishery.

The Vanuatu Cultural Centre has in recent years embarked on the Vanuatu Young People's Project that helps urban and rural youth to document their lives, analyze their situation and voice their opinions. It would seem that this type of project would be useful in rural communities such as Lelepa where young people could be empowered to contribute to resource management and other village institutions.

There is at present little communication between men and women, and also between youth and their elders, on topics related to the fisheries and resource management. Before women and men, elders and youth can co-operate actively in local resource management, people will have to abandon certain unhelpful gender stereotypes. Some men paint women as mindless and destructive harvesters, whereas some women portray men as selfish and unwilling to listen. Also, traditional norms stifle women's public voices.

Men, women and youth all need to be recognized as legitimate stakeholders who have something positive to offer. Opening up the dialogue may seem dangerous to some because it contravenes traditional cultural norms. Custom is not, however, a static edifice. Culture is a living and mutable social arrangement and as such it is open to negotiation. In Melanesia especially, traditional institutions have "a virtually unlimited capacity for accommodating new things" (Hviding 1998: 255). Gender and generational roles, which influence the shape and success of resource management arrangements, have been changing steadily through time and will continue to change. The challenge is to identify and promote the types of change that will allow for sound and equitable resource management.

Main street through Lelepa's villages.

NOTES

1. At the time, 100 vatu was equivalent to US $0.907.
2. VCC = Vanuatu Cultural Centre. The centre has volunteer field workers who live and work in rural villages throughout Vanuatu.

ACKNOWLEDGEMENTS

This research was supported by a grant from the Canada-South Pacific Ocean Development Program. We also wish to thank the University of the South Pacific Marine Studies Programme and Vanuatu Cultural Centre for their support. Thanks to Douglas Kalotiti for assistance in the field, to Francis Hickey for help with illustrations, and most of all to the people of Lelepa who took care of us and gave generously of their time to answer our questions.

REFERENCES

Hickey, F. R. and R. E. Johannes 2002. Recent evolution of village-based marine resource management in Vanuatu. *SPC Traditional Marine Resource Management and Knowledge Information Bulletin* 14, October, pp.8-21.

Hughes, T. P. & J. H. Connell, 1999. Multiple stressors on coral reefs: a long-term perspective. *Limnology and Oceanography*, 44:932-940.

Hviding, E. 1998. Contextual flexibility: present status and future of customary marine tenure in Solomon Islands. *Ocean and Coastal Management* 40:253-269.

Johannes, R. E. 1998. Government-supported, village-based management of marine resources in Vanuatu. *Ocean and Coastal Management* 40:165-186.

National Statistics Office, 2000. *The 1999 Vanuatu National Population and Housing Census - Main Report.*

Pyle, R. L. 1993. Marine aquarium fish. In A. Wright & L. Hill (eds) *Nearshore Marine Resources of the South Pacific*. Suva: IPS, FFA & ICOD. pp.135-176.

14 Gender Collaboration: A Case Study of Local Resource Management in Safa'i Village, Samoa

Malama Jasmine Siamomua-Momoemausu

"E au le ina'ilau a tina."
Women are capable of doing all things.

— This is a traditional Samoan saying often heard when a woman achieves something that is beyond expectations, especially the expectations of the opposite sex.

GENDER ROLES IN FISHERIES MANAGEMENT

The conservation and management of natural resources has been a central issue in international fora, for example at the United Nations Conference on Environment and Development (UNCED) of 1992 and other succeeding conventions. Agenda 21 stresses the importance to indigenous peoples of the marine environment and of community-based subsistence fisheries. It points out the need to include traditional ecological knowledge and socio-cultural values as inputs to management (United Nations 1993). Chapter 17 of Agenda 21, which provides guidelines and strategies for sustainable development of fisheries resources, acknowledges the importance of being inclusive of all stakeholders who have a significant role in fisheries including, in particular, women.

Women in coastal communities across the Pacific have been involved from time immemorial in many aspects of fisheries (e.g. harvesting, processing and marketing) but the role of women in fisheries has often been undervalued or neglected in government plans and processes (Shoeffel 1985, Matthews 1993, Steele 1993, Tuara 1995). Slatter (1994) and Tuara (1995) proposed that women are neglected because governments focus not on subsistence inshore fisheries, in which women predominate, but on large-scale industrial fisheries that mostly involve men. Although inclusion of women in fisheries development has been promoted in the several last decades they have nevertheless been excluded from many development planning processes (Lal & Slatter 1982, Poussard 1993, Ram 1993). Even

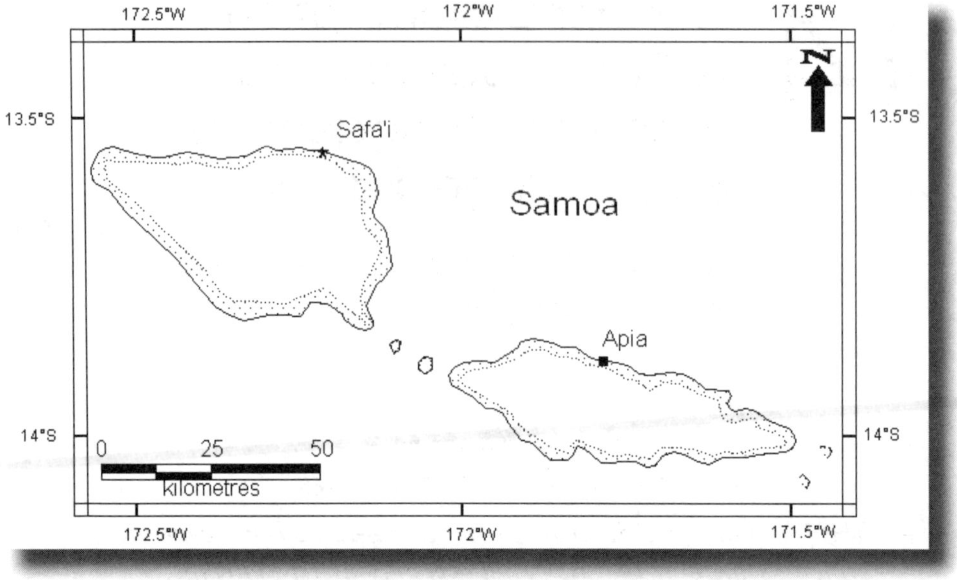

Map 1. Location of Safa'i on the island of Savai'i in Samoa.

at the village level women, along with other marginalized groups, may be left out from development projects and decision making when male chiefs dominate these (Ram-Bidesi 1998).

Fisherwomen deserve to be heard in the decision making process because, like their male counterparts, they are important users of marine resources. They should be assigned a particularly key role in management and conservation of inshore stocks (Taniera & Mitchell 1992). The exclusion of women means that their knowledge of inshore habitats and resources is not used, nor are their attitudes and opinions made clear; this may lead to the failure of a community-based fisheries project (Steele 1993). The solution to the problem rests in the hands of national fisheries authorities. It is only when governments invest in village fisheries and work at the grassroots level, following an inclusive approach, that both men and women will be fully involved in fisheries management and conservation.

There is limited published information concerning inshore fisheries in Samoa. This study explores gender roles in fisheries management and conservation, focusing on the involvement of women and men of the village of Safa'i in the development of a village level Fisheries Management Plan (FMP) in 2002. In Samoa in recent years, the Department of Fisheries has taken care to include women in their fisheries management and conservation programs. Women are daily subsistence harvesters of marine resources and are therefore immediately affected by every decision and intervention that is made. The inclusive approach has proven extremely important for the long-term success of these activities. This case study shows how women in Safa'i were involved in developing their Fisheries Management Plan. I also include an assessment of some impacts of the Plan on the role of women in the fishery and offer points to consider in the quest for making women more fruitfully involved in fisheries management activities.

CASE STUDY METHODS

Interviews with the Fisheries Division staff responsible for the Fisheries Extension Program in Safa'i took place in the main office in Apia (Upolu Island) and the sub-office at Salelologa (Savai'i Island) in early October 2003. This was almost one year after the village's fisheries management plan was developed. Information regarding people's perceptions of male and female participation in fisheries management and conservation was drawn from interviews with Safa'i village chiefs, key male fishers and members of the village women's committee. Additional information was acquired from personal observations of the village and from focus group discussions with key members of the Safa'i Fisheries Management Committee (women and men fishers). In addition, previous publications on fisheries and women were consulted.

THE EVOLUTION OF VILLAGE-BASED FISHERIES MANAGEMENT

Safa'i Village and Traditional Management Practices

Safa'i is a small, traditional community located on the northern side of Savai'i Island (Map 1), the bigger of the two main islands of Samoa. It is home to an estimated population of 250 people excluding villagers who have migrated to New Zealand, Australia and other places. Agricultural and fishing activities, which are conducted at both subsistence and artisanal levels, are the mainstays of the village economy.

The main fishing methods include spear fishing, bottom fishing and netting. These are performed mainly by men, for commercial purposes. Women and their children glean shellfish, sea cucumbers and other species from the reef flats for family consumption and at times for income generation. According to village informants, fishing was not effectively regulated in the past and this resulted in the decline of some fish species. The countrywide rise in the standard of living has impelled the drive for cash income and consequently caused people to exploit the inshore marine resources beyond sustainable levels. Two cyclones that struck the islands in 1990 and 1991 also caused severe damage to the marine environment, particularly to the coral reefs.

In the past, the marine resources in Safa'i were managed through traditional rules and measures that every villager was expected to observe. As in all other rural villages in Samoa, rules concerning fisheries were devised, implemented and enforced by male chiefly authority. Most rules involved bans on various destructive fishing practices such as dynamite, poisons and coral smashing. People from other villages were prohibited from fishing in the waters of Safa'i.

When fisheries rules were developed in the past, women were excluded from the formal consultation and decision-making process; they waited for their husbands to inform them of what had gone on in the men's meetings. Even though they were excluded from developing the rules, women were expected to abide by them.

The Transition to Cooperative Management

The traditional management measures, which were administered by chiefly decree in the past, failed to work effectively because of the lack of full community support and poor enforcement. People continued to disobey traditional rules as their demand for cash incomes from selling fish accelerated. This pressure was exacerbated by the heavy concentration of human population in coastal areas. Faced with degradation of the marine environment and the decline of fisheries resources, the government recognized that it could not carry out management alone. In response to research that showed the decline in catches of inshore fish and shellfish (Horsman & Mulipola 1995, Zann 1991) the government introduced a Fisheries Extension Program in 1995. Over the subsequent years this program has gradually devolved powers, roles and responsibilities for inshore fisheries management back to the local communities. The approach was to improve management at the community level through getting communities involved in the development of their own Fisheries Management Plans (FMPs) for the use of marine resources. In August 2002 Safa'i participated in this program and went on to approve its Plan two months later.

The Process of Building a Cooperative Management Institution

In the early stages of the Program, three target groups were consulted: women (the Women's Committees, *tina* and *tamaitai*), titled men (*matai*) and untitled men. In separate meetings, each group analyzed the problems in their marine environment such as the degree of change that had occurred in seafood catches. They identified the various causes of the problems and proposed possible practical solutions.

The main concerns and solutions that were noted by the various working groups are shown in Table 1. The idea to establish a Marine Protected Area was one of the key requests made by village groups as a solution to declining fish stocks.

The Current Approach: Cooperative Management

The cooperative management approach used by the Fisheries Extension Program has established an institutional structure for the control of marine resources that enables collaborative actions between the village *matai* system and the government. As government-imposed fisheries regulations were and are mostly ineffective without community support, communities have been given the primary responsibility of managing their marine environment and its resources. As stressed by King and O'Sullivan (1996): "The

Table 1: Main concerns and proposed solutions revealed during group discussions.

Concern	Solution
Too much rubbish on the beaches and in the sea	Enforce village rules and have regular beautification activities
Illegal fishing by neighboring villages	Establish village bylaws
Unregulated fishing of undersized fish and shellfish	Enforce related government and community fisheries regulations
Use of destructive fishing methods including the use of SCUBA diving	Ban the use of such methods
Declining fish stocks	Establish Marine Protected Areas Monitor compliance with village rules Restock lagoons with shellfish supplied by Fisheries Division

Source: Modified from Anon 2002.

Government recognizes that fish stocks cannot be effectively protected simply by passing national laws. The Village Fisheries Extension Program is different in that it acknowledges that the real solution to the problem lies in the hands of the village people and their *fonos*".

The program is sustained through shared decision-making. In the formulation of the Fisheries Management Plan, every effort was made to include both women and men in all stages of the process (King & O'Sullivan, 1996). At the time of this study, the Plan had been in place for one year and Safa'i people were very optimistic concerning their fisheries management efforts. Both men and women were observed to be active in the daily monitoring of their protected areas on land and in the sea.

Members of the Women's Committee were elected to sit with male representatives in the Fisheries Management Advisory Committee (FMAC) which further considered the problems and solutions previously identified. This committee decided how the solutions could be implemented with the support of both management parties, and determined the penalties for violations. With the assistance of the Fisheries Division staff, the Committee prepared a draft Fisheries Management Plan, which was further discussed and approved by the village *fono*.

Both female and male FMAC members were appointed by the village *fono* to an implementation body, the Fisheries Management Committee (FMC). This committee was and is responsible for the monitoring and enforcement of the Fisheries Management Plan.

Elements of the FMP

The Plan is an agreement between the village and the government. It sets out on one hand, the community's undertakings for marine resource management and conservation, and on the other hand, the service and technical support to be undertaken by the Fisheries Division (King & Fa'asili, 1997). Some of the key undertakings of Safa'i's Fisheries Management Plan that have major impacts on women include the following:

- Bans on the use of destructive methods (e.g. plant poisons (*derris* roots), chemicals and dynamite, coral smashing).
- Enforcement of fish size regulations and other existing village rules.
- Establishment of a Marine Protected Area (no fishing area) in part of the traditional fishing ground.
- Establishment of Fisheries bylaws to guide fishing activities.
- Restocking of lagoon and reef areas with giant clams and Trochus.
- Implementation of other fisheries development projects (e.g. fish and turtle fences).
- Workshops and training for community members on aspects of fisheries management and conservation.

During group discussions in Safa'i, both men and women reported that their opinions contributed much to the design of management measures. When identifying the cause and effect of problems, the villagers had been willing to face the fact that the problems in the marine environment were largely of their own making. They recognized the impacts of their own destructive fishing activities.

The village declared a 200m long stretch of its adjacent, traditional inshore fishing ground a Marine Protected Area (MPA) and devised a range of conservation undertakings that were subsequently implemented. At the time of this study, the MPA had been free from exploitation for one year. How long it remains untouched is totally up to the village *fono* (council of chiefs) to decide.

Village strategies for generating alternative sources of income and ensuring a supply of seafood included the implementation of two fisheries projects in their inner lagoon. (1) The fish fence project nurses and reserves fish (e.g. mullet) that are thought to be rare in the village waters. These fish either migrate from the sea at high tide or are spotted and captured by the men during open water fishing. Fish are kept penned until the village chiefs declare a day of harvest. This is usually a special occasion where the fish will be served to the *fono* or shared among all families of the village. (2) The turtle conservation project keeps turtles that the fishermen involuntarily capture when fishing. Turtles are kept within a secure fence, and this serves as an income-earning tourist attraction[1] that contributes significantly to the development of the village. It was widely commented that families no longer have to donate money for village activities as this is taken care of by the income generated from this project.

Walking the Talk

A key ingredient in the success of the programs in Safa'i has been that everyone shares the work and responsibilities of inshore fisheries management. Personal observations in Safa'i revealed the involvement of both women and men in fisheries management activities. Women contribute to the upgrading and maintenance of turtle and fish farms by raising funds (e.g. through concerts, food stalls) to pay for building materials such as wires, nets and cement. In some instances, women also help the men in erecting and cementing stone fences and walkways within the reserves and along the shore in front of the village.

Collecting the rubbish on the beach and tree planting at the village's beachfront are done only by women. When, with the assistance of the Fisheries Division, a television documentary was produced to promote Safa'i's fisheries management and conservation projects, it was a woman who was featured prominently in the film. When asked why, she answered with a smile *"Because I had managed to do it right"*. Participation in management activities has become a point of pride for the women involved.

In Samoa, women are well regarded as advisers of families and communities. Besides being the main providers of family food, women contribute to socio-economic stability, improving everyone's nutritional and living standards both through their domestic chores and through income generating activities that supplement family incomes. Women may not speak out in public but behind the scenes, they contribute greatly to the stability and effectiveness of the village *fono*. Their inclusion in the various FMP working committees speaks of the men's recognition and support of their roles in society. In interviews, male respondents admitted that to avoid disaster, it is essential to work with women on fisheries programs and activities. Women were said to contribute to monitoring and enforcement of the Fisheries Management Plan by either telling off trespassers or reporting them to the chiefs.

The situation in Safa'i has provided an opportunity to identify strategies that aim at improving the participation of women in fisheries management. To do so we must first explore the impacts of the new management arrangement on women, and understand the motivations they have to comply or non-comply with the new restrictions.

IMPACT OF THE FISHERIES MANAGEMENT PLAN ON WOMEN

Abandonment of Lucrative but Destructive Fishing Practices

Women generally use very simple but sometimes also destructive fishing methods and gears. The interview respondents included a woman whose favorite means of catching small fish used to involve discharging poisonous plant material *(derris* root*)* into the water. By refusing destructive methods, women also no longer target under-sized fish. Destructive fishing methods in some cases were key economic activities. For example, one woman who had been consistently fishing sea anemones by breaking and removing corals recognized that her activities had been destructive and senseless. She admitted, "My involvement in this management program has taught me a very vital lesson. I quit [fishing anemones] completely though my family has no more staple income source. My relatives outside always urge me but I refuse…conservation now matters more to me than my greed". There is every reason to believe that women are now more conservation-minded than before because of having the Fisheries Management Plan in place.

A Change of Fishing Space and Time Allocation

The implementation of Safa'i's Fisheries Management Plan has significantly affected women's fishing. The damaged fishing grounds most needing management were the inshore areas on which women were heavily dependent. Turning part or all of the inshore into a 'no fishing' zone forces a paradigm shift in women's social and economic roles. After the MPA was set up, each woman had to decide whether or not to continue to fish.

To the women, the establishment of the Marine Protected Area spelled a change of fishing ground from one that was close to home and shallow to one that was further away and deeper. They even changed from employing very simple fishing methods to adopting the fishing gears and techniques typically used by men.

Attached to the spatial change has been an increased allocation of time for fishing. The further away the fishing ground, the more time is taken to reach it. Women's fishing trips now take much longer than before, which means they spend more time away from home duties of cooking and minding the children.

A Change in Family Nutrition

As a result of the MPA, some women shifted from daily fishing to part time. For some this means going to sea only once in two weeks or not at all. This has resulted in a change in the family's protein source from fresh fish to imported, tinned fish and meat, a change that may gradually affect the health of family members.

A Change in Domestic Routine

Fishing is no longer a daily routine for most women and when asked how things had changed for them, this was the most prevalent response. About 90 percent of the women interviewed said that after the MPA was set up they turned more to land-based activities to earn income and provide food for their families. This means that the land resources on which women rely are now more heavily exploited.

As well as participating in the village's land-based management activities as required under the Plan, women are increasingly engaged in the development of small family businesses to supplement family food and income. These include selling home-made food products and handicrafts, growing vegetables and fruits and rearing poultry and small livestock.

Social Problems

Faced with the need to travel out to deep water to fish, women sometimes decide to break accepted norms by fishing from the inshore areas of neighboring villages. Such a move may create disputes between the villages. Although taking this risk indicates a commendable degree of support for their own village rules, it also indicates a need for further education concerning consequences of shifting fishing pressure from one inshore area to another.

Hope for the Future

The long-term goal of establishing the Marine Protected Area, lagoon and reef restocking, enforcement of fisheries by-laws, rules and regulations is to improve future catches of fish and shellfish and ensure sustainability of fisheries. Interviews revealed that within the first few months after the plan was enacted, signs of increasing abundance of fish and shellfish were observed in the village's reef and lagoon areas (Anon 2002). Women will directly benefit from the long-term improvements that will result from sacrificing their fishing space. They will no longer have to spend a lot of time gleaning and going farther out in search of a day's catch. Most importantly, having the Management Plan in place ensures sensible and sustainable exploitation of inshore resources.

In order to continue supporting the new management rules, women have had to adjust themselves to many changes. They must face the associated challenges for as long as the area remains a 'no fishing' zone. Even so, it is not likely that unsustainable fishing will follow the lifting of the ban. With their increased awareness of resource conservation, women now speak positively about sensible exploitation in future when the ban is lifted.

ENHANCING THE PARTICIPATION OF WOMEN IN FISHERIES MANAGEMENT

Women in Fisheries Workshops and Training

One goal of the Program was to gain the maximum participation of women. More women are taken on board whenever opportunities become available for fisheries related workshops and training. Women express appreciation for their inclusion in the various stages of the Program; it has motivated them to take the Program seriously. Providing training for women has been a significant move for the Fisheries Division in Samoa towards achieving sustainable management and development of marine resources.

Cultural Impediments to Participation of Women

Interviews revealed that cultural barriers keep women from participating in some village activities. This is not seen to be a problem for most women, who hold traditional beliefs and are satisfied with their way of life, including having to bear both the reproductive and productive responsibilities for the home. This responsibility is especially heavy where single women are the key supporters of children and elder family members. Government and village leaders therefore need to recognize women's realities. For example, women may have less free time than men for attending meetings. Meetings may have to be scheduled at times that accommodate their domestic routines. If women have appropriate support they can reach their full potential as collaborators in fisheries management.

BUILDING ON THE SAFA'I EXPERIENCE

Throughout its history, Safa'i has been a mere village; one would never know it existed in Samoa. Its participation in fisheries cooperative management since 2002 has brought gradual changes for the better to the village fisheries. Safa'i is now very successful and therefore widely known in Samoa. This case study has documented one of the key factors in Safa'i's success: the active participation and collaboration of both women and men.

Both fisheries staff and the village chiefs responsible for cooperative management in Safa'i affirmed the significant involvement of women. Women helped in the development of the Plan and, subsequently, in sustaining marine resources. The Plan has affected women's fishing activities, social and economic roles. They had to cope with negative impacts and changes to their lifestyle that followed implementation. It is possible that if they had not been involved in the planning, women would have been more resistant to these changes and to the new rules.

The study shows what can happen in villages that are participating in the Fisheries Extension Program in Samoa. The Fisheries Management Plan has provided a reliable way for enforcing village rules and regulations that were already in existence. It has brought villagers together to support the village *fono*, making it more firm and effective. In spite of the potential difficulties for women inherent in the new rules, the stability of the village *fono* and of the women's committee provided an open door for the government to negotiate and initiate new management strategies. The women's committee also made it possible to implement strategies that aimed at making women more fruitfully involved.

It is not appropriate to assume that every village will respond to the program in the same way as Safa'i. Different villages have different social structures, and they may not devise the same undertakings and fisheries projects as those of Safa'i. However, some general lessons can be learned which could be applied in other villages in future. For example, one lesson learned is that the major part of the task of devising and carrying out a successful Plan falls on the village people collectively. Therefore, it makes sense to be as inclusive as possible to maximize the work force. For this to work optimally, there is a need for some adjustments to the traditional approach which privileges males in decision-making processes and tends to provide information and training primarily to males. Conscious adjustments are needed, for example in planning the timing and place of meetings so that women can attend. These are changes that people at all levels must approve.

Another general finding is that equipping women by providing education, training and awareness are fundamental ways of making them realize their abilities and potential.

Some recommendations for consideration are:

1. The Fisheries Division should continue to strengthen communication links with village level women's committees and encourage their involvement in fisheries workshops and training.
2. Once they have been made fully aware of the causes of declining fisheries and the effects of local management, the women's committees should be used as an avenue to promote fisheries management in and around the villages.
3. Data on the subsistence fisheries and/or the involvement of women in fisheries in Samoa are vital for resource assessments, policy planning, etc. but are limited. Through better communication, women can provide or be assisted to acquire such needed information.
4. The Fisheries Division should undertake periodic reviews of Fisheries Management Plans to determine the progress of women and to further encourage their direct involvement in the implementation and monitoring phases.
5. The capacity of the national government to assist fisherwomen is limited. Women should therefore take the initiative to set up a 'Women in Fisheries Association' in Samoa and seek assistance and support from other women's institutions in the Pacific region such as the Women in Fisheries Networks in New Caledonia and Fiji.

From this case study I conclude that, if given a chance, women can confidently make decisions and collaborate productively with men in tasks that have up to now been assumed to be a male domain. If women had not been included in the process of developing Village Fisheries Management Plans, their knowledge and concerns over the inshore marine resources may not have been heard or included in the plan, and compliance with rules as well as community participation in activities may well have suffered. An inclusive process has, in the case of Safa'i, resulted in a strong and hopefully resilient local management arrangement. At all levels of fisheries management, the collaboration of both women and men should be encouraged and supported as a basis for achieving the goal of sustainable fisheries.

NOTE

1. Admission charge is WST 4 per head. At the time of writing(2003), one Samoan tala (WST) was worth US $ 0.36.

ACKNOWLEDGEMENT

To the Assistant Director of Fisheries and staff of Samoa Fisheries Division - a big *faafetai* for the support rendered during the course of this study. God bless you all.

REFERENCES

Anon. 2002. *Safa'i Fisheries Management Plan.* Samoa: Fisheries Division, Ministry of Agriculture, Forests, Fisheries & Meteorology.

Horsman, N. & A. Mulipola 1995. *Catch Data and Collection from Market Surveys in Western Samoa.* Joint FFA/SPC workshop on the Management of South Pacific Inshore Fisheries. Noumea, New Caledonia: SPC.

King, M. & U. Fa'asili 1997. *Community-Based Management of Fisheries and the Marine Environment.* Pacific Science Association Inter-Congress, Suva, Fiji. July, 1997.

King, M. & N. O'Sullivan 1996. *Village Fisheries Extension Programme.* Information Sheet 1. Western Samoa: Fisheries Division, Ministry of Agriculture, Forests, Fisheries & Meteorology.

Lal, P. & C. Slatter 1982. *The Integration of Women in Fisheries Development in Fiji.* Report on an ESCAP/FAO initiated project on improving the socio-economic condition of women in fisherfolk communities. Suva, Fiji: Fiji Fisheries Division and the University of the South Pacific.

Matthews, E. 1993. *Women and Fishing in Traditional Pacific Islands Culture.* Workshop on People, Society and Pacific Islands Fisheries Development and Management, Selected Papers. Inshore Fisheries Resources Project Technical Document. Noumea: South Pacific Commission, p.29 – 33.

Poussard, W. 1993. Of women and fish. SPPF. *Tok Blong* 44.

Ram, V. 1993. Women in commercial fisheries in the South Pacific. In G. R. South, *Marine Resources and Development.* The Ray Parkinson Memorial Lectures. Suva, Fiji: PIMRIS, University of the South Pacific, p.105-119.

Ram-Bidesi, V. 1998. Inshore fisheries management regimes in small island economies of the South Pacific: problems and prospects for a sustainable fishery. Paper presented at the Islands of the World V: International Conference on "Small Islands in the Third Millennium – problems and prospects of island living", Le' Reduit, Mauritius, 1-5 July, 1998. Suva, Fiji: University of the South Pacific.

Shoeffel, P. 1985. Women in the fisheries of the South Pacific. In *Women in Development in the South Pacific: Barriers and Opportunities.* Canberra, Australia: Development Studies Centre, Australian National University.

Slatter, C. 1994. For food or foreign exchange? Regional interest versus global imperatives in Pacific Fisheries development. In 'Atu Emberson-Bain, *Sustainable Development or malignant growth? Perspectives of Pacific Island Women.* Suva, Fiji: Marama Publications, p.131-148.

Steele, S. 1993. Women in fisheries development. Workshop on people, society and Pacific Islands fisheries development and management. *Inshore Fisheries Research Project Technical Document* No. 5. Noumea, New Caledonia.

Taniera, T. & J. Mitchell 1992. Notes from Kiribati (August 1992). In E. Matthews, *Fishing for Answers, Women and Fisheries in the Pacific Islands.* Suva, Fiji: Women and Fisheries Network, p.29-32

Tuara, P. 1995. *The Role of Women in the Management of Pacific Island Inshore Fisheries.* FFA/SPC Workshop on the Management of South Pacific Inshore Fisheries. 26 June – 7 July, 1995. Noumea, New Caledonia.

United Nations 1993. *Programme of Action for Sustainable Development,* the final text of agreements negotiated by governments at the United Nations Conference on Environment and Development, 3-14 June 1992. New York: UN, p.117-165.

Zann, L. 1991. *The Inshore Resources of Upolu, Western Samoa: Coastal Inventory and Fisheries Database.* Field Report No.5. FAO/UNDP Project SAM89/002.

15 Gender In Pacific Maritime Communities: Putting the Social Into Science and Connecting the Local to the Global

Vina Ram-Bidesi and Jean Mitchell

INTRODUCTION

This volume of case studies privileges the research of Pacific Islanders and their documentation of the wide range of marine-related activities that are part of everyday life in Oceania. It provides descriptions of the complex ways in which islanders around the Pacific interact with their maritime environments. The goal is to stimulate more sustainable use and management of the limited natural resources of the Pacific so that these resources can meet the growing and changing needs and aspirations of Pacific peoples.

There is great diversity within the Pacific. A wide range of highly variable cultural, environmental and historical factors have shaped the contemporary conditions under which individuals and communities live and participate in local networks (Lockwood 2004). While there are many differences, all Pacific Islanders are connected by the vast and bountiful Pacific Ocean which provides sustenance, cash and the site for social relations, symbolic meanings and cultural practices. The complexity of the marine world of Islanders is reflected in Nero's idea that "Islanders' lived-in world continues beyond the beach into, through and over the seas, the underwater terrain of the lagoon and sea-routes to neighboring Islands as familiar to Pacific fisher as fields and roads to a farmer" (1997:374). Connections are an important theme in this volume: connections that relate communities to their ecological domains; tradition to modernity; subsistence to the cash economy; men to women and the local to the global. Oppositional and dichotomous terms often used when discussing these relationships have often diminished our understanding of their complexities and mutability.

There are profound changes occurring in Pacific coastal communities. This volume of case studies is an important contribution for it provides insight into how coastal resource use and management strategies have evolved in response to the changing social, cultural, technological and economic realities that are being negotiated by coastal dwellers. Case study results indicate that not only do development and change threaten ecological sustainability in some communities, but economic development is not always socially desirable. Under such circumstances, economic gains from development may be short-lived, providing private benefits to only a few individuals and promoting livelihoods that are not sustainable. There are also case studies in this volume that provide evidence of a balanced approach to coastal fisheries management and development. It is important to ask what lessons can be drawn from these divergent experiences in order to ensure the future well-being of communities.

PUTTING THE SOCIAL INTO SCIENCE

Many research methods based on the unexamined western concepts of fisheries science fail to consider the social and cultural dimensions of fishing. This volume underlines the necessity of connecting the scientific to the social in order to acquire a comprehensive understanding of complex inshore fisheries. Many would agree that effective programs and sustainable management strategies hinge on people and the changing ways in which they are reckoning the world. Without combining social and biological approaches to reach an understanding of the marine world there can be little success in management and little hope for sustainability, no matter how good the science. This volume attempts to address the invisibility of the social aspects of fishing and in particular, the lack of attention paid to the gender dimensions of inshore activities.

Woman displays crabs for sale in Suva market, Fiji

THE NECESSITY OF GENDER ANALYSIS

Connecting social to scientific information means that we draw attention to the realm of human action, which is always gendered. Gender is one of the key organizing themes in this volume and it provides the analytical framework necessary for probing the nuanced relationships and complex changes that occur in households and communities. Gender analysis provides a way to understand social categories and symbolic meanings as well as material realities in coastal societies.

This volume emphasizes the different ways in which men and women utilize and understand their maritime environments. Analysis does not simply segregate women and men into respective domestic and political spheres. Attention to the details reveals that the division of labor is not simply a dual gender construct. Gender roles encompass the attention and effort that women and men give to production, whether the activities are shared or separate. Gender analysis also explores the ideas and practices of men and women that mediate the tensions inherent in social life (Weiner 1992:4).

This use of a gender perspective springs from the recognition that women and men who make their living from the inshore are often absent from key decision-making processes affecting their fishery and its management and development. Studies included in this volume attest to the importance of disaggregating information based on gender. This process has allowed the researchers to account for social and economic costs and benefits to both genders, making it possible to fairly assess the sustainability of fisheries and suggest appropriate measures for future management and development.

This volume demonstrates the practical realities of changing fisheries practices and the ever-pressing challenges of management and conservation. More importantly, it provides insights as to why consideration of gender is essential if sustainable fisheries are to be achieved for the well being of everyone in the coastal population. Key findings illustrate that women and men often differ in their roles and responsibilities within their households and communities. The information and knowledge they hold therefore differ and their perceptions about marine resource use and management are not necessarily the same. An examination of gender in the context of the inshore fishery also raises a number of important concerns regarding women's work and women's status, the effects of development policies and the consequences of integration into global markets.

THE VALUE OF WOMEN'S WORK

Paying attention to gender illuminates the broad scope of women's productivity and cultural activities and underlines the need for closer attention to women's production, particularly where it is seen to be ephemeral and valueless (Weiner 1992:x). Women's fishing activities, which are concentrated on inshore resources in both rural and urban areas, encompass fishing for home consumption, the marketing of fish, post harvest processing of fish in villages or factories and production of exchange items such as shell products for tourism and for cultural purposes. All of these marine related activities are enmeshed in the changing political economy of the many atolls and islands of Oceania.

SHIFTING GENDER ROLES

As various contributions in this volume clearly demonstrate, gender as a social and cultural arrangement is subject to change (e.g. Fidali-Hickey & Whippey-Morris, this volume). The boundaries between men's and women's fishing are shifting. In some areas, women fish beyond the immediate shore areas and are participating in what were previously defined as men's fishing activities, to take advantage of commercial opportunities (Vunisea, this volume). In other cases, women who fished close to home are fishing further away because some resources have become depleted (Siamomua-Momoemausu, this volume, Tarisesei & Novaczek, this volume). Elsewhere, women now actively fish where previously they hardly fished at all (Agassi, this volume). The introduction of new technologies such as nylon fishing lines, gill nets, spear guns and dynamite shape gender roles in the fishery. Technologies also affect the allocation of time to different fisheries as well as the utilization of particular species.

OVERFISHING AND LOSS OF FISHERIES HABITAT

Many studies, including those concerning Vanuatu, Tonga, Tuvalu, Solomon Islands, Kiribati and Samoa, documented how fishing grounds in close proximity to the human settlements were over-exploited and damaged because of increased fishing pressure and the use of destructive fishing methods. Subsistence fishing activities are often considered informal and therefore insignificant but with increased demand for food and income, they can intensify to the point of doing serious damage. Impacts of fishing may be further exacerbated by impacts of other coastal development activities that either compete within the same coastal space or cause destruction to fisheries habitats.

Because their fishing activities are seen as marginal, women and children are rarely considered as potential players for implementing resource management and conservation. The underestimation of the contribution of women's fisheries to household nutrition also leaves women without a voice when it comes to coastal development activities. The social losses that follow when subsistence fisheries are negatively affected by coastal development are often ignored, leaving women, children and others involved in this type of fishery with little or no compensation.

In response to localised areas of over-exploitation and habitat damage, women and children have to travel further in search of productive fishing grounds. This may impose additional costs for investment in gear. It also either leaves women with less time to do household chores and to attend to social responsibilities or extends women's working hours, putting stress on their physical health.

WOMEN IN DECISION-MAKING

Many case studies, for example those dealing with the shell fisheries in Tuvalu (Resture & Resture, this volume) and Kiribati (Fay-Sauni & Sauni, this volume) and the *Trochus* fishery in Tonga (Malimali, this volume) illustrate that even where women are dominant harvesters, there is no institutional framework that integrates them into the decision making process in their respective fisheries. Often these women lack critical information on issues that affect the resources they target or their access to those resources. As a result, women may have little incentive to comply with resource conservation measures (Tarisesei & Novaczek, this volume) or development decisions (Tekanene, this volume) and as a result these program or project objectives are not fully achieved. In the case of Safa'i village in Samoa (Siamomua-Momoemausu, this volume), women and men from the village were consciously consulted and subsequently contributed to management decisions. Consequently, the fisheries management strategies have worked well, with positive results. Thus we see that valuing people's knowledge and understanding their practices is vital to the success of participatory, community based management of the inshore.

Many case studies identify the subordinate positions of women as well as youth in coastal communities. Women and young people are a crucial part of the fishing force, but within hierarchical structures, male elders make key decisions on behalf of the community at large. Often, the assumption is that such decisions are in the best interest of all people whom the leaders represent. When a village elder or the chief does not consult with the people and does not share the economic benefits of development, disillusionment with traditional decision-making processes develops (Agassi, this volume, Tarisesei & Novaczek, this volume). Consequently, resource management rules may be ignored and often, social tensions increased. To resolve such situations and to bring about effective resource use and management systems, it is necessary for women and young people to have a voice in framing community projects. Such projects also need to be thoroughly evaluated using a gender sensitive feasibility analysis.

Rural woman selling fish laplap in Honiara city market, Solomon Islands.

THE LOGIC OF THE MARKET AND THE EROSION OF SELF-RELIANCE

For many islanders, the cash and subsistence and formal and informal sectors are interwoven. When islanders move in and out of market production and subsistence production facilitates this process, they are insulated to some degree from the vagaries of the market (Rodman 1993). This ability underpins self-reliance in many Pacific communities. The separation of subsistence and cash economies was introduced during colonialism, perpetuated by postcolonial policies and has been intensified by globalization. In everyday life, as revealed in the case studies, island households undertake occupational plurality that involves both cash and subsistence activities and straddles both informal and formal sectors. Being flexible in the allocation of their labor and maximizing subsistence strategies have long been important dimensions of women's work. Only by maintaining a variety of potential sources of income, food and social support can rural islanders ensure continued self-reliance. Thus, there are households where one will see some kind of fishing, fish being sold, fish being given away, fish being eaten and fish being processed. In addition, some members of the household will be working in an urban centre or overseas, sending remittances whilw dreaming of their island home.

Schoeffel (1992) has pointed out the long-standing notion that the subsistence sector can continue to meet the needs of islanders and accommodate changing population and market demands. The profound dependence on the subsistence sector, of which the inshore fishery is part, needs to be emphasized (Sauni & Fay-Sauni, this volume). The new stresses and old strains on this sector, which is now so heavily interrelated with the cash economy, were illuminated by many authors in this volume (Vunisea, this volume, Fidali-Hickey & Whippy-Morris, this volume, Tekanene, this volume, Resture & Resture, this volume). There is evidence that market opportunities are upsetting the always delicate balance between meeting subsistence needs and generating cash. Women may take their fish to market or abandon fishing in favor of wage labour and then use the cash earned to purchase nutritionally inferior tinned fish and meats. As Vunisea (this volume) has noted, proximity to markets changes the time and effort allocated to fishing (at the expense of other activities) and also affects the distribution of catches. Target species are determined increasingly by the market's preferences. In the past, people fished for food and exchanged the surplus in various ways that supported community relationships. Today the catch is increasingly marketed; sometimes none is saved for home or exchange.

As the changes from increased globalization (i.e. cash and the goods and services that can be purchased) trickle down to the community, people who come to depend on purchased food are placed at a higher risk of losing their self-reliant food production strategies and the traditional knowledge attached to these strategies. Subsequently, nutritional vulnerability of the community is increased.

Declining self-reliance combined with a lack of alternative opportunities to earn money can result in poverty not only in urban areas (e.g. Levuka in Fiji, Tarawa in Kiribati, Funafuti in Tuvalu) but also, as seen in the case of the Solomon Islands shell fishery, in rural areas. The shortage of food in some households and a surplus of the "wrong kind" of foods such as tinned fish and meats, sugar and white bread have resulted in a rapid rise of non-communicable diseases such as diabetes and heart disease. The new range of health problems and household food insecurity have emerged as formidable problems for many nations of the Pacific in the postcolonial era.

Domestic as well as international markets effect changes in small-scale fishing patterns and local self-reliance. Case studies in Fiji, Solomon Islands and Tuvalu provide examples of artisanal/commercial fisheries that service industrial and export markets with, for example, bait fish and non-edible marine products including aquarium fish, shells and coral. These new fisheries provide economic opportunities in coastal communities but they also create new socio-economic dynamics as fishers neglect subsistence fishing in an attempt to maximize cash incomes. As noted by Sauni et al (this volume), the apparent improvement in financial well-being may be illusory, especially in the long term. In addition, the ecology and vulnerabilities of such new fishery targets are not adequately known. In the event of over-exploitation and the collapse of such a fishery, people may be left with few resources and traditional skills to support future livelihoods.

Caption: Smoking sea cucumbers in rural Fiji to make bêche de mer for the international market.

CONNECTIONS BETWEEN GLOBAL AND LOCAL

Hau'ofa (1993), in his important article, reminds us that location and perspective matter. In Oceania there are two different points of view operating at very different levels. At the macro level there are politicians, bureaucrats, experts and development planners who are devising plans, policies and development programs. At the micro level there are people, whom Hau'ofa suggests are living their lives in imaginative ways, making a living and caring for their families and communities. These people who occupy the micro world and, for our purposes, the inshore zone, go about their everyday lives knowing important things about the world in which they live. They capitalize on new opportunities, recognize the impact of change and face the constraints imposed by changing socio-economic processes. The creativity of human agency and, as this volume demonstrates, the importance of gender considerations, are all too often left out of economistic analyses used at the macro level. Analysts who focus on standard economic indicators often provide bleak descriptions of development in the Pacific that emphasize structural problems and inherent limitations. One consequence of such an approach is that the inshore and the subsistence fishery are devalued because they do not generate large amounts of cash. According to Adams, Dalzell and Ledua (1999), Pacific Island fishing vessels harvest only a small fraction (about 7%) of the total catch in the Pacific region. The large commercial/industrial fishing fleets exploiting fish resources of the Pacific

Ocean are mostly purse seiners from distant fishing nations: the United States, Japan, Taiwan and Korea. The fishing rights sold to these foreign fleets constitute a major source of income for island nations. The focus in this volume on the inshore fisheries of the Pacific is particularly important, because the inshore is where most islanders fish for consumption and for domestic markets.

The world wide expansion of capitalism that has resulted in the dominance of distant water fishing nations in the Pacific fisheries operates through free trade and neo-liberal practices and drives global flows of capital and technologies. Globalization promotes the integration of national economies. The concomitant extension of market systems means that goods, information, people and ideologies are moving around the world in large quantities and at record speeds (Lockwood 2004). Case studies in this volume show how new market demands, consumption patterns and technologies have become part of life in some Pacific communities.

It is important to examine the intersection of the local and global and to understand the new ways in which men and women and their communities participate in this globalizing world. To varying degrees, the processes of modernity, e.g. colonialism and the introduction of capitalism, have shaped the islands and atolls of the Pacific and their inhabitants. They continue to be shaped by what we now call globalization. People have been and continue to be drawn into the world system through a process of resource extraction and participation in capitalist markets as fishers, copra producers and migrant laborers. Some of the intersections of local and global are demonstrated in this volume. Examples include the collapse of the copra market, the growing need for cash, the decline of marine fish stocks due to excessive harvesting, migration for wage jobs, consumption of imported tinned fish and meat, women's employment in fish processing plants, and the move away from collective concerns to individual aspirations.

THE GENDER OF GLOBALIZATION

The experiences and consequences of the processes of modernity and globalization have been reshaping the ideas, practices, and relationships among and between women and men. Trade liberalization is one aspect of globalization that affects women, households and communities. This is particularly evident in Rajan's (this volume) research in Fiji, which shows how market pressures affect the working conditions for women in a fish processing plant. Multinational companies such as Bumble Bee take advantage of the vulnerability of companies such as PAFCO in Fiji. In turn, PAFCO uses a preference for female labour and the imposition of inferior working conditions and low wages to maintain its competitive edge. Women, who make up the bulk of PAFCO workers, may work for years as "temporary" laborers and with token wage increments, if any. By keeping a large number of women in casual positions, the company avoids paying benefits that would accrue to established staff. Performance standards are set at such high levels that women find it almost impossible to gain any promotion or wage increment.

Most tuna canneries and processing factories are located in developing countries where employment options for people are limited (see for example Bonanno & Constance, 1996). Women are drawn to such employment but find that chances for progress are minimal. In rural settings, female workers perform their factory work plus domestic responsibilities that extend their hours of physical labour, ultimately affecting their health and productivity. There little community response or corporate concern over these issues. Women who cannot maintain the required pace are simply branded as being lazy or incompetent.

Kiribati fishmongers selling tuna by the roadside in Tarawa.

TRADITION AND MODERNITY

The relationship between tradition and modernity is far more complex than the typical oppositional constructions would suggest. The marine world of the Pacific is resonant with cultural and social meanings that have not been erased by the logic of the marketplace. For example, this volume demonstrates the cultural significance of shellcrafts in Tuvalu and the Solomon Islands. At the same time, women's incomes gained from the crafting of shells are essential to their families. In Solomon Islands, demand for shell money has increased despite the move to cash economies; in Tuvalu markets for shells are also expanding. Thus are traditional cultural objects shaped by market demands and socio-economic changes.

Reciprocal exchanges have long been at the center of Pacific sociality and women have made an essential contribution to social production by participating in local food exchanges (Tekanene, this volume). Exchanges are most often reciprocal among kin and neighbors and they facilitate and maintain kinship and social networks (Veitayaki & Vesi, this volume). Although fish surpluses are now often sold, that does not mean that important social exchanges are ended. Yet, as seen in this volume, inequities are emerging at household and community levels. There is less willingness to share resources and collective activities have been eroded as people pay more attention to their short term economic interests. Even though social security is assured through kinship networks these may, in modern times, be seen to be at odds with the growing need to do business and to fulfil individual needs (Fidali-Hickey & Whippy-Morris, this volume).

With reference to traditional management systems, case studies also reveal tensions about traditional leadership arrangements. Nevertheless, traditional culture provides important frameworks for managing

Women selling giant clam meats in the Suva market, Fiji. In many parts of the Pacific these shellfish are classed as endangered species.

resources. Hviding (1998) has noted the "contextual flexibility" at the heart of *kastom* or traditional practices in Melanesia, which means that these systems can accommodate changes and modern pressures.

The importance of traditional knowledge and practices and their transmission to the younger generations is widely acknowledged. The inter-generational transmission of local knowledge becomes more difficult as young people and their families move to urban areas or migrate to other countries in search of jobs and modern services.

DEMOGRAPHIC SHIFTS: MIGRATION, URBANIZATION AND YOUTH

Demographic shifts including migration to towns and the youthfulness of many Pacific island populations are important trends that were mentioned in a number of the contributions to this volume. In 20 out of 22 Pacific countries the median age is under 25; at least 11 have median ages below 20 years (SPC 2004). Finding employment is often highly problematic for young people, yet youth expect that wage labour will be available if they complete their education and move to an urban area.

Migration to town and to other destinations outside of the country, such as is occurring in Tuvalu, Samoa and Tonga, has many implications. Where a majority of young men have migrated from a community, as in Nukufetau in Tuvalu, domestic economies are profoundly affected and women must assume increased responsibilities for family welfare (Resture & Resture, this volume). When migrants into towns fail to find paid employment they often revert to rural patterns of subsistence involving fishing. Peri-urban, inshore fishing grounds such as those near Port Vila, Honiara and Tarawa are coming under increasing pressure and represent some of the most degraded of marine environments. Because this is where many young people are growing up, such areas deserve attention so that the fishing grounds can be restored and properly managed. This represents an extra challenge because traditional management approaches are less likely to succeed in culturally heterogeneous, urban societies. Not surprisingly, the case study from Vanuatu, with its youthful and largely underemployed population, makes the case for inclusion of young people in decision-making processes that will shape their futures. Both urban and rural youth must be factored into fisheries development, conservation, management and knowledge production. Since 1997 the Vanuatu Cultural Centre has actively included young people in its programs and has advocated nationally for the needs of youth.

Migration has meant the concentration of populations in capital cities, for example in Fiji, Tuvalu and Kiribati. Women's work in marketing fish in such urban areas is very important for the distribution of catches and the improvement of diets of urban dwellers. And yet, women who are fish-mongers have little meaningful support for their important marketing work (Tekanene, this volume, Vunisea, this volume).

CREATING ALTERNATIVE RURAL LIVELIHOODS

The need for more diverse and sustainable income generating opportunities in rural communities was a theme in many of the case studies. Economists generally hold that a country or community should specialize and trade in a commodity where they have some comparative advantage. Several authors drew our attention to the collapse of the copra market in rural areas of the Pacific. This activity, upon which many rural islanders depended for a small but reliable source of income, is an example where specialization resulted in vulnerability. The collapse of world copra prices led inevitably to a shift of pressure to the inshore fisheries as a source of income. The inshore is now suffering the consequences.

While it is often argued that the production systems of Pacific Islands lack economies of scale and often use inefficient technologies, there is scope for the creation of diverse and high value products for sale in niche markets. For example, products from well-managed fisheries can command a premium price in the world market through eco-labeling (Phillips et al 2003). Women and youth who target commodities such as aquarium fish, corals and shells and who farm products such as pearls and seaweeds can also potentially serve niche markets by maintaining high quality production and marketing systems. First, however, it must be firmly established that there are no long term, negative ecological implications of such trades as aquarium fish, corals and shells.

In the case of the shell trade, women currently arrange their own markets and promote their products with very limited if any assistance from community leaders or Fisheries Departments. This is because shell collecting is not recognized as a commercial fishery. The potential for the improved marketing of shell handicrafts requires investigation. If broader markets can be reached, women should be assisted through training to participate in resource management and the sustainable use of shells, and to optimize the economic potential of the trade.

SUMMING UP

Global processes, the policies developed by governments and the economic decisions taken at the macro level have tangible effects on communities throughout the Pacific. Impacts of development choices on the environment and on society can be dramatic and the costs are often disproportionately borne by women and the young. The theme of forfeiting the future for short-term gains permeates this volume of case studies, suggesting the need for a fresh approach to fisheries development and management. More gender sensitive research into small scale and subsistence fisheries is needed. Approaches put forward by authors in this volume are ones that value local knowledge and traditional institutions and recognize the complexity of relationships that underpin both the commercial and subsistence activities of communities. Case study authors also call for an expanded dialogue that creates spaces for women and young people in development processes and in resource management. These voices are too often silenced despite the fact that women and youth are key players in inshore fisheries.

Consideration of both social and biological information is important not least because of the depth of cultural attachment Pacific islanders feel for the sea. As Vunisea (this volume) and Malimali (this volume) have pointed out, the social and cultural dimensions of fishing are integral to life in coastal communities. These insights come from islanders who have experienced everyday life in the Pacific: such perspectives are the strength of this volume. People go to the inshore fishing grounds not just because they need to; they want to go for the sheer enjoyment of being near the water and for the pleasure of sharing time with others. These rich and relatively intangible values add urgency to the search for appropriate and effective solutions to the problems of degradation of inshore resources. The ocean is a place where social relationships are forged, social obligations fulfilled and local knowledge transmitted. It is a place to embrace the world of Oceania.

REFERENCES

Adams, T., P. Dalzell & E. Ledua 1999. Ocean resources. In M. Rapaport (ed) *Pacific Islands: Environment and Society*. Honolulu: Bess Press. P.366-381.

Bonanno, A. & D. Constance 1996. *Caught in the net: the global tuna industry, environmentalism, and the state*. USA: University Press of Kansas.

Hau'ofa E. 1993. Our sea of islands. In *A New Oceania: Rediscovering Our Sea of Islands*. Suva: USP\SSED

Hviding, E. 1998. Contextual flexibility: present status and future of customary marine tenure in Solomon Islands. *Ocean and Coastal Mangement* 40:253-269.

Lockwood, V. 2004. The global imperative and Pacific Island societies. In V. S. Lockwood (ed) *Globalization and Culture Change in the Pacific Islands*. New Jersey: Prentice Hall.

Nero, K. 1997. The material world remade. In D. Denoon (ed) with S. Firth, J. Linnekin, M. Meliesea & K. Nero, *The Cambridge History of The Pacific Islanders*. Cambridge: Cambridge University Press.

Phillips, B., T. Ward & C. Chaffee (eds.) 2003. *Eco-labelling in Fisheries. What is it all about?* England: Blackwell Publishing.

Rodman, M. 1989. *Deep Water: Development and Change in Pacific Village Fisheries*. Boulder: Westview.

Rodman 1993. Keeping options open: Copra and fish in rural Vanuatu. In V. Lockwood (ed), *Contemporary Pacific Societies*. New Jersey: Prentice Hall.

Schoeffel, P. 1992. Women in Fisheries in the Pacific Islands: A Retrospective Analysis. Unpublished.

SPC 2004. Pacific Island Populations 2004. www.spc.int/demog

Weiner, A. 1992. *Inalienable Possessions*. Berkeley: University of California Press.

About the Authors

Alan Sepe Agassi

Alan Agassi was born beside the Marovo Lagoon in Western Solomon Islands. In 1991, he completed a Graduate Diploma in Urban and Regional Planning at Queensland University of Technology, followed in 2000 by a Postgraduate Diploma in Development Studies at the University of the South Pacific. Currently Alan works as Chief Rural Development Officer in the Ministry of Provincial Government & Rural Development of Solomon Islands. He is also the Executive Chairman of an NGO partner to the Ministry, The Rural Development Volunteers Association.

Lilian Fay-Sauni

Lilian Fay-Sauni was born and lived most of her life in Kiribati. She graduated with a BAppSc (Fisheries) from Australia and an MSc (Marine Science) from the University of the South Pacific. Lilian has worked in Kiribati, Tuvalu and Fiji. Her jobs have included Senior Fisheries Officer, Tutor/Demonstrator and Research Assistant at USP, and Coordinator for the Women in Fisheries Network. Currently she does consulting work with the Secretariat of the Pacific Community, and is residing in Noumea.

Kristina Fidali-Hickie

Kristina Fidali-Hickie comes from Malaita Province in the Solomon Islands. A teacher by profession, Kristina graduated in 1999 with a Bachelor in Education (Biology) degree from the University of the South Pacific, Fiji. She worked as a Fisheries Officer from 2000-2002, promoting marine awareness. At the end of 2002, Kristina joined a new environmental NGO, ECANSI, as a Project Officer. Her interest lies in researching the strength of women and whether that strength can be harnessed to meet the challenges of conservation in the marine environment of the Pacific. Kristina is married and a mother to one child.

Mecki Kronen

Mecki Kronen is of German origin. She holds a PhD in geography and a master's degree in marine sciences. At present Mecki works as socio-economist in the European Union funded PROCFish/C project that is implemented by the Secretariat of the Pacific Community's (SPC) Reef Fisheries Observatory. Her work focuses on documenting the demography, social and economic structures, and subsistence and small-scale artisanal fisheries (finfish and invertebrate) of the South Pacific.

Pierre Labrosse

French fisheries scientist Pierre Labrosse was the Head of the Reef Fisheries Observatory of the Secretariat of the Pacific Community from 1998 to 2003. After gaining a PhD in tropical ichthyology at the University of Perpignan (EPHE), Pierre went on to survey inshore fisheries and fish habitats and to design multi-disciplinary research and technical assistance programmes on reef fisheries systems across the Pacific region. He is currently based in Senegal, where he is involved in the design of a Fisheries Knowledge Exchange System for western Africa.

Siola'a Malimali

Siola'a Malimali, born in Tonga, is a fisheries officer who works in the aquaculture division of the Tonga Ministry of Fisheries. He graduated from University of the South Pacific with a B.Sc. in Marine Science in 1991 and would like one day to continue on with graduate studies in fisheries.

Jean Mitchell

Jean Mitchell, born in Prince Edward Island, Canada, is now teaching anthropology at the University of Prince Edward Island. After completing an MA in Development Studies Jean worked extensively overseas. She later gained her MA and PhD in Social Anthropology from York University in Canada. Since 1991 Jean has conducted research in Kiribati on women's fishing. In Solomon Islands she has written two situation analyses of women and children. In Vanuatu she has researched migration, urbanization and postcoloniality and started the Vanuatu Young People's Project with the Vanuatu Cultural Centre.

Irené Novaczek

Irené Novaczek was born in Scotland and has lived in Canada most of her life. She is a research scientist, environmental activist, small business entrepreneur and mother. Irené gained her PhD in marine ecology from the University of Auckland, New Zealand and lived in eastern Indonesia for several years before starting to work on small scale fisheries and small business development in the South Pacific in 1999. She is currently Director of the Institute of Island Studies at the University of Prince Edward Island.

Jyotishma Rajan

Jyotishma Rajan was born and brought up in Rakiraki, Fiji and did a degree in Environmental Studies at USP, winning the gold medal and Shell Fiji prize for outstanding performance. Her research interests include the impacts of globalization on small developing countries and communities, environmental management and conservation, and gender equity. She completed her Postgraduate Diploma in Marine Affairs in 2001 and is now finishing her Master's program. Jyotishma currently works with Greenpeace as the Energy Campaigner. One of the focus areas of her work is ensuring that women especially in rural areas have access to safe, reliable and non-polluting sources of renewable energy.

Vina Ram-Bidesi

Vina Ram-Bidesi is a lecturer at the Marine Studies Programme of the University of the South Pacific. Her research interests focus on fisheries management and development but she also has keen interest in the areas of trade and environment. Vina has worked on both community-based inshore fisheries and offshore fisheries, as well as economic and policy analysis, gender issues and integrated resource management. She has been a member of the Board of Trustees of the Women and Fisheries Network, an NGO advocating gender-sensitive research in the fisheries sector.

Alan Resture

Alan Resture, originally from Tuvalu, is a Fellow at the Institute of Marine Resources of the University of the South Pacific, working on curriculum for Distant and Flexible Learning. In the past Alan served as an Assistant Lecturer at the Marine Affairs Programme of the USP. He has a background in Marine Affairs and Geography, and holds a Masters of Urban and Regional Planning from the University of Hawai'i. Alan's research interests are in the area of community-based resource management, socio-economic impacts of climate change, and coastal area planning for small island states.

Setapu Resture

Setapu Resture, born in Tuvalu is a secondary school teacher by profession. She holds a Bachelor of Education (Chemistry) degree from the University of the South Pacific, Fiji and is currently enrolled in a Bachelor of Arts degree (Mathematics/Economics). She hopes to continue her research on women in fisheries activities.

Samasoni Sauni

Samasoni Sauni comes from Nukufetau atoll in Tuvalu. He gained his BAppSc (Fisheries) from Australia and an MSc (Tropical Coastal Management) from England. Samasoni has served as senior fisheries officer in Tuvalu, as executive director of the influential Fijian environmental NGO, SPACHEE, and as USP Project Coordinator of the Canada-South Pacific Overseas Development (C-SPOD) regional project in post harvest fisheries development. He is currently working for the Secretariat of the Pacific Community (SPC), based in Noumea. His interests and area of expertise are reef fisheries ecology and coastal management. Samasoni's doctoral research is on localized and small-sized taboo areas and coral bleaching, primarily assessing their combined effects on reef fish communities.

Malama Siamomua-Momoemausu

Malama Siamomua- Momoemausu was born in Samoa and gained her BSc and Postgraduate Diploma at University of the South Pacific. She has been a permanent Fisheries Officer of Samoa Fisheries Division since 1996 and is currently on study leave to pursue a Masters Degree in Marine Studies at the USP. Her thesis topic concerns village fishing and the role of women, areas of work that are yet to be studied in depth in Samoa.

Jean Tarisesei

Jean Tarisesei comes from Ambae Island in the north of Vanuatu. She works in the Vanuatu Cultural Centre as a Women's Culture Project Coordinator, where she is responsible for women's cultural affairs. For her work Jean travels to many islands to talk to women about the culture and traditions of Vanuatu. While working on a Women's Traditional Marine Tenure Project, she collected information about traditional knowledge of near shore marine resources in Vanuatu and developed a video on the topic, called Reef Survivors. The video is in use for public education throughout Vanuatu. Jean is also the mother of five children.

Maere Tekanene

Maere Tearikua Tekanene was born on Tabiteuea North in southern Kiribati. After earning a Diploma of Education from USP and teaching for many years, Maere went to Victoria University of Wellington for a Diploma in the Teaching of English as a Second Language. She subsequently attended James Cook University, where she graduated with a Bachelor of Arts. At the end of 2001 Maere stopped teaching to work with the Kiribati Council of Women and the Pacific Foundation for the Advancement of Women, but she continues to assist parents operate a school for young drop outs and youth unable to secure secondary schooling on South Tarawa. Her work in community facilitation includes action research, documentation of the implementation of the Beijing Platform of Action and the Status of Women in Kiribati, and the assessment of resources, gaps and best practices in science and technology. Maere is married to Tenanora (Rera) Tekanene and mother to four children.

Joeli Veitayaki

Joeli Veitayaki of Fiji is the Coordinator of the University of the South Pacific's Marine Affairs Programme and Director of the International Ocean Institute-Pacific Islands. Joeli's research interests are in human ecology. Together with other practitioners from Fiji, he is working to involve communities in the effective management of their marine resources and to encourage others to do the same in other Pacific countries. Joeli has published widely in the areas of customary marine tenure, capacity building, marine resources management and regional cooperation. He is currently writing his second book based on rural development and fisheries development projects.

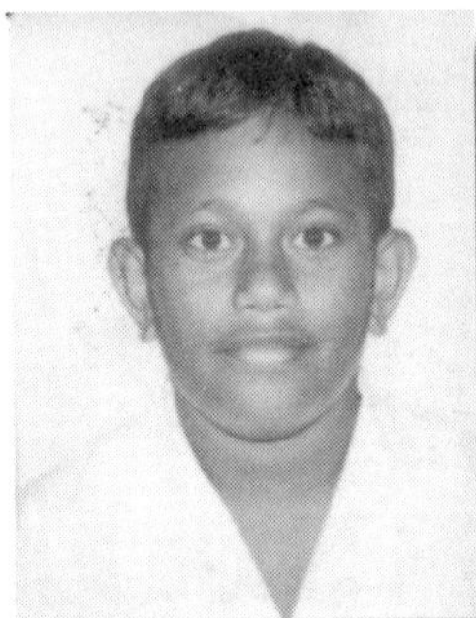

Waisake Paul Vesi

Waisake Paul Vesi is an 11 year old Fijian boy attending Yat Sen Primary School in Suva. During the Christmas holiday in 2003, Vesi went to his village and witnessed the annual fish drive on Malawai. He found himself taking pictures of the fish drive and learned some of the fishes' names. He even managed to catch one fish!

Aliti Vunisea

Aliti Vunisea is currently the Secretariat of the Pacific Community's Coastal Fisheries Management Officer, assisting SPC member countries to put together management plans, conduct socio-economic surveys in coastal fisheries communities and promote participation of women in all fisheries sectors. Aliti has conducted studies on current and potential gender participation in the tuna industry as a component of the FFA National Tuna Development and Management Plans. Before joining SPC, she worked as a lecturer at the University of the South Pacific, where she completed a Masters degree on "Women in Subsistence Fisheries in Fiji". She also worked as community trainer for the International Ocean's Institute, coordinated the Women in Fisheries Network and acted as a resource person for NGOs and communities interested in coastal fisheries management, women and gender participation.

Cherie Whippy-Morris

Cherie Whippey-Morris comes from Fiji and has had considerable experience working on community-based projects. Through her work with NGOs and in the Fisheries Department, Cherie has gained insight into regional, national and community fisheries issues. Her particular interests are women in fisheries, aquaculture, and the intersection of gender and environment. In future, Cherie would like to make a significant contribution to the training and education of Pacific fisher-women.

Index